R&D PRIORITIES IN INNOVATION POLICY AND FINANCING IN FORMER SOCIALIST COUNTRIES

NATO Science Series

A series presenting the results of scientific meetings supported under the NATO Science Programme.

The series is published by IOS Press and Springer Science and Business Media in conjunction with the NATO Public Diplomacy Division.

Sub-Series

I.	Life and Behavioural Sciences	IOS Press
II.	Mathematics, Physics and Chemistry	Springer Science and Business Media
III.	Computer and Systems Sciences	IOS Press
IV.	Earth and Environmental Sciences	Springer Science and Business Media
V.	Science and Technology Policy	IOS Press

The NATO Science Series continues the series of books published formerly as the NATO ASI Series.

The NATO Science Programme offers support for collaboration in civil science between scientists of countries of the Euro-Atlantic Partnership Council. The types of scientific meeting generally supported are "Advanced Study Institutes" and "Advanced Research Workshops", although other types of meeting are supported from time to time. The NATO Science Series collects together the results of these meetings. The meetings are co-organized by scientists from NATO countries and scientists from NATO's Partner countries – countries of the CIS and Central and Eastern Europe.

Advanced Study Institutes are high-level tutorial courses offering in-depth study of latest advances in a field.
Advanced Research Workshops are expert meetings aimed at critical assessment of a field, and identification of directions for future action.

As a consequence of the restructuring of the NATO Science Programme in 1999, the NATO Science Series has been re-organized and there are currently five sub-series as noted above. Please consult the following web sites for information on previous volumes published in the series, as well as details of earlier sub-series:

http://www.nato.int/science
http://www.springeronline.nl
http://www.iospress.nl
http://www.wtv-books.de/nato_pco.htm

Series V: Science and Technology Policy - Vol. 46 ISSN 1387-6708

R&D Priorities in Innovation Policy and Financing in Former Socialist Countries

Edited by

Walter Leal Filho
TuTech Innovation, Hamburg, Germany

and

Plamen Svetoslavov Gramatikov
Southwest University "Neofit Rilsky", Blagoevgrad, Bulgaria

IOS Press

Amsterdam • Berlin • Oxford • Tokyo • Washington, DC

Published in cooperation with NATO Public Diplomacy Division

Proceedings of the NATO Advanced Research Workshop on
R&D Priorities in Innovation Policy and Financing in Former Socialist Countries
5-8 May 2004
Blagoevgrad, Bulgaria

ISBN 1 58603 478 2
Library of Congress Control Number: 2004115806

Publisher
IOS Press
Nieuwe Hemweg 6B
1013 BG Amsterdam
Netherlands
fax: +31 20 620 3419
e-mail: order@iospress.nl

Distributor in the UK and Ireland
IOS Press/Lavis Marketing
73 Lime Walk
Headington
Oxford OX3 7AD
England
fax: +44 1865 750079

Distributor in the USA and Canada
IOS Press, Inc.
4502 Rachael Manor Drive
Fairfax, VA 22032
USA
fax: +1 703 323 3668
e-mail: iosbooks@iospress.com

PRINTED IN THE NETHERLANDS

Introduction

While the nations of Central and Eastern Europe transform their economies from a planning and centralised control towards free market systems, they face the need to formulate their R&D priorities and to change their systems of innovation. The problems of restructuring the R&D and innovation systems of the post-socialist countries have so far been treated as marginal within the overall institutional transformations that are taking place in these nations.

In other words, innovation and in particular, support mechanisms to innovation have not been a priority in the past.

It is a fact that progress in science, technology, and innovation-related policy-making in the transitional countries has been rather modest and has largely failed to keep pace with other areas of social, political, and economic change. Most of the transitional countries lack the institutional capability for sophisticated policy making on R&D, for technology transfer and innovation and for building consensus among experts, corporate decision makers and members of the research community on R&D and innovation policy issues.

There is relatively little recognition of the growing importance of knowledge and continuous life-long learning as the basis for economic success in the 21st Century. One of the tools that may help the modernisation of R&D and innovation-related policy-making is the training and retraining of policy-makers and knowledge and innovative managers. Moreover, it is important to equip them with the tools to take advantage of the various financing possibilities available for innovation.

This book, which contains a set of the papers delivered at the NATO Advanced Research Workshop on "R&D Priorities in Innovation Policy and Financing in Former Socialist Countries" held in Blagoevgrad, Bulgaria in May 2004, attempts to explore the different dimensions of innovation and how it relates to R&D development and financing in a set of eastern and central European countries. The matter of funding for innovation (especially, but not only EU support) is both sensitive and important, and is considered in this book.

First of all, this publication addresses the need to clarify the matters related to the application and awards of such funding. It is a fact known by well-informed people that most EU funding is not paid directly by the European Commission – as many unfortunately believe – but via the national and regional authorities of the Member States. This is the case for payments, for example, under the Common Agricultural Policy and for most payments under the structural policy financial instruments (e.g. European Regional Development Fund, European Social Fund, European Agricultural Guidance and Guarantee Fund, Financial Instrument for Fisheries Guidance) which make up, in money terms, the great bulk of EU funding.

Secondly, it contains analyses and appraisals which show the extent to which innovation is being practised in a number of central and eastern European countries (e.g. Bulgaria, Lithuania, Poland, Russia, Ukraine) and the difficulties associated with it. At the same time, some of the papers show that some progress is indeed being achieved in many areas.

It is believed that the combination of the information provided by the authors and the case studies presented by them will provide a concrete support for efforts towards strengthening the financial basis of R&D and innovation in post-socialist countries.

Walter Leal Filho (Hamburg) and *Plamen Svetoslavov Gramatikov* (Blagoevgrad)

Contents

Chapter 1
Current priorities in R&D and Innovation Policy in Post-Socialist Countries: Results of a Survey

WALTER LEAL FILHO
TuTech Innovation
Harburger Schlossstrasse 6-10
21079 Hamburg
Germany

Summary

This chapter outlines come of the current priorities in R&D in post-socialist countries and reports on the results of a survey of innovation in a set of countries. It also includes an analysis of elements related to financing and some of the needs to be met to pursue innovation in a more sustainable way.

Introduction

The current emphasis on innovation as a source of industrial competitiveness, and hence prosperity, became especially noticeable in Europe in the early 1990s. A series of brainstorming meetings on innovation and technology diffusion policy had been held prior to drawing up the Maastricht memorandum "An Integrated Approach to European Innovation and Technology Diffusion Policy" in 1993, which stated that "innovation and technology policy has an important role to play in developing the competitiveness of industry in the face of global competition and in improving the quality of life. The latter causes an essential need to guide the process of technological change towards important public goals such as environmentally sustainable development" (European Commission, 1993). Further, the European Commission's 1995 Green Paper on Innovation and the subsequent First Action Plan for Innovation in Europe were launched and provided an additional momentum (European Commission, 1995).

Despite the above developments, a matter of concern is the fact is that even for highly developed countries in Europe, a major bottleneck in the research-development-innovation system is the use of research results for market exploitation. This shows that we are still a long way from fully responding to the economic and social challenges of innovation in the 21st century – which is meant to be the century of innovation.

It is widely acknowledged that innovation is the key factor in industrial competitiveness for any country. In order to help research achieve concrete outcomes and economically beneficial and socially desirable results, mechanisms need to be set up to promote innovation, exploit the results of scientific work, and stimulate the creation of innovative companies.

According to Viesturs and Draveniece (1998) many countries in eastern and central Europe, including Latvia, have already passed through a phase of the transition process. Now they have to face the subsequent phase of integration of their economies into the world economy and markets, which calls for the modernization of industry and technology policies in accordance with world trends. Nowadays, most of the world´s developed economies are knowledge-based and consequently a good future for all countries, in Europe as well as in the rest of the world, will depend on their willingness to give priority to the development of knowledge and investments in human capital.

It is believed that the main trends in the knowledge-based economies will also occur in the less developed countries. Therefore it can be expected that in C&E Europe the role of the higher education and R&D (or RTD) sector will change dramatically by promoting them into a core component of the national economy. These create a stock of knowledge and human capital which constitutes the main production and investment factors in knowledge-based economies.

The development of a democratic society for countries in transition requires efforts in the following areas:

(a) To establish an adequate higher education sector; achieve international recognition of higher education, i.e. the recognition of awarded degrees;

(b) To maintain and enhance the research potential of the laboratories, universities and research institutes (centres) and their ability to produce knowledge at the highest level (publications in journals, participation in international scientific conferences, work abroad, etc.) in the areas which comply with educational programs and other local priorities. In other words, to help ensure that research serves the state's economic and social objectives;

(c) To ensure that innovation processes and the technology transfer process in both fundamental research and applied research lead to real technological developments and to the full implementation of their results.

There is no other way to new technologies than through education that is both permanent and suited to new technological requirements, and innovation. The secret lies in knowing how to convert scientific progress into industrial and commercial success, and traditional institutions like universities also have to retrain themselves. There is nothing in common here with the craze for innovation *per se*, since the cardinal function of new technologies is not to recreate a paradise on earth but rather to mitigate – often at a huge cost – the towering difficulties of future survival.

Can C&E European countries innovate?

The answer to this question, posed by Högselius (2003) is a clear "yes". However, there are various factors that influence innovation, such as macro-economic reforms, trade liberalisation, and enterprise privatisation, and economic performance in terms of GDP growth, productivity, and trade. These variables are, of course, important, but in the globalising learning economy, it may be even more important to assess reforms related to systems of innovation, and to study performance in terms of innovative strength.

As a result, one can legitimately ask the question to what extent firms and other organisations in Central and Eastern European countries do have the ability to innovate at all. Whereas from the perspective of GDP growth C&E European countries can in general be seen to be 'catching up' with the West, there is a widening gap between East and West with respect to the underlying issue of innovative output, suggesting that C&E European countries are 'falling behind'. An interpretation given by Högselius is that transition to capitalism has so far been associated with decreasing rather than increasing innovative capabilities, and that the eastern European economies are becoming locked into a pattern of growth based on low value-added activities related to the exploitation of low-wage labour. If this is true, Högselius goes on, it would be an alarming result from the perspective of C&E European countries governments, since public policies in these countries contain a strong commitment precisely to the establishment of advanced innovation-driven economies with strong systems of innovation and R&D bases, where education, science and technology would represent the main driving forces of socio-economic change.

A further question C&E European countries may be asked is how best to promote innovation? One way of doing so is by finding the proper balance of competition and patent law and policy. Although both competition in markets and patents for inventors can work together to foster innovation in western Europe, either policy requires a proper balance with the other to do so. Consumers and innovators win when patents and competition policy are aligned in the proper balance. Although questionable patents can harm competition and innovation, valid patents work well with competition to promote innovation. This means that C&E European countries need to find ways for patent systems to maintain the proper balance with competition.

In this context, legislative and regulatory changes are needed. Policies of questionable validity can slow further innovation and raise the costs for the consumer. Therefore in order fully to capitalise on innovation, there is a need to create new administrative procedures that will make it easier for firms to innovate with few expensive and time-consuming barriers.

The European Trend Chart on Innovation: results of a survey

Innovation is a priority of all Member States and of the European Commission. Throughout Europe, hundreds of policy measures and support schemes aimed at innovation have been implemented or are under preparation. The diversity of these measures and schemes reflects the diversity of the framework conditions, cultural preferences and political priorities in the Member States.

The "First Action Plan for Innovation in Europe", launched by the European Commission in 1996, provided for the first time a common analytical and political framework for innovation policy in Europe. Building upon the Action Plan, the "Trend Chart on Innovation in Europe" is a practical tool for innovation policy-makers and scheme managers in Europe. Run by the Innovation Directorate of DG Enterprise, it pursues the collection, regular updating and analysis of information on innovation policies at national and Community level, with a focus on innovation finance, the setting up and development of innovative businesses, the protection of intellectual property rights and the transfer of technology between research and industry.

The "Trend Chart on Innovation" (European Commission – Directorate-General Enterprise, 2003) serves the open policy coordination approach laid down by the Lisbon Council in March 2000. It supports policy-makers and scheme managers in Europe with summarised information and statistics on innovation policies, performances and trends in the European Union. It is also a European forum for benchmarking and the exchange of "good practices" in the area of innovation policy.

The Trend Chart on Innovation has been running since January 2000 (European Commission – Directorate-General Enterprise, 2003). It tracks innovation policy developments in all EU Member States, plus Bulgaria, Cyprus, Czech Republic, Estonia, Hungary, Latvia, Liechtenstein, Lithuania, Norway, Poland, Romania, Slovak Republic and Slovenia. The Trend Chart website provides access to the following services and publications, as they become available:

- a database of policy measures across Europe;
- a 'Who's who?' of agencies and government departments involved in innovation;
- a series of six-monthly country reports for all countries covered;
- a series of six-monthly trend reports covered on each of the four main themes;
- a number of benchmarking reports;
- the European Innovation Scoreboard and other statistical reports;
- a news service and thematic papers;
- the annual reports of the Trend Chart.

There have been many interesting developments in respect of innovation in C&E European countries, many of which have gone largely unnoticed. In this paper, we present the results of a survey aimed at identifying, documenting and reporting on such trends. The main information source for this survey was the country reports produced in the context of the "European Trend Chart on Innovation" ((European Commission – Directorate-General Enterprise, 2003 a, b, c, d, e, f, g, h, i), to which information this section is substantially indebted. It is believed that a detailed look at what happens in each country allows not only a general view of current developments, but also the identifications of gaps and differences. To this purpose, progress in a sample of C&E European countries , which includes Bulgaria, Estonia, Hungary, Latvia, Lithuania, Poland, Romania and Slovakia has been reviewed, and will be looked at in detail on a country per country basis.

a) Bulgaria

New firm creation, and more particularly the creation of firms with an advanced technological base (NTBFs) and openness to global markets, are particularly important factors in the restructuring the Bulgarian economy. The political changes in the 1990s allowed start-ups to flourish in the country. A process for improving the conditions for new firm creation, particularly for hi-tech ones, is now taking place. NTBFs are supported through the amendments to the Law on Standardisation, which removes conflicts of interest between standardisation and regulatory functions and/or certification. Considerable efforts are still needed to reinforce administrative procedures for standardisation and certification and to create systems for national conformity assessment and market surveillance (notably laboratory upgrading and accreditation). Company law has also been further aligned but still needs to be brought fully into line with elements of the *acquis*, especially on acquisitions, mergers and divisions of companies.

Several measures have been taken by the government to stimulate the creation of new firms, and particularly high-tech ones. However, the lack of policy tradition in this area creates difficulties for the creation of an adequate portfolio of measures. The national strategy for hi-tech development adopted by the government in 1999 aims at increasing the role of hi-tech in the contemporary economy and at improving the conditions for the development of hi-tech activities in the country. The current policy is based on the implementation of this document and includes actions for improving the business climate in the country, developing industrial zones, plans for innovation strategy etc.

There have been initiatives such as the *Guarantee fund for Micro-crediting* project (BG 03), which started at the beginning of 2001 in 18 pilot municipalities with the highest rates of unemployment. This fund has been created with money from the state budget, with an initial capital of €10 million. The objective of the project is to create new jobs through facilitating the access to funds for small and newly created enterprises, for people willing to establish their own businesses, and to free the resources necessary for the development and widening of business activity. There are two types of loans: investment loans of up to €7,653 and working capital loans of up to € 5,102. Terms of payment are 5 years for investment loans with a six months grace period, and 18 months for working capital loans with a one month grace period. The Guarantee Fund had extended loans for 616 projects at a total value of €3.06 million by the end of September 2002. Forty-two percent are dedicated to start-up companies. The Fund guarantees 70% for SMEs and 100% for start-ups, and annual interest rates are at 10%. Since 1 August 2002 this Fund has been applicable to all municipalities in Bulgaria.

There is a significant shift in the nature of the links between firms and research bases during the transition period. These institutional changes have negatively affected knowledge transfer and as a result the position of Bulgaria in respect to the technology transfer index is very low. In a table of other transition economies and neighbouring countries, Poland is 6th, Hungary is 8th, Turkey is 15th, the Czech Republic is 24th, Greece is 32nd, Slovakia is 37th, and Bulgaria is 56th.

An example of cooperation between industry and universities is the consortium between the Bulgarian Industrial Association and 8 Bulgarian universities, which has now been operating for five years. The consortium's activities seek to achieve higher competitiveness for companies and to maintain this competitiveness in a dynamically developing, economic environment, as well as to increase staff competitiveness and skills through continued training in compliance with the requirements of international markets and consumer demands.

Partnerships between Bulgarian and foreign universities are a very important component of entrepreneurial training. Primary projects were implemented under the Tempus and PHARE programmes. Departments delivering training in entrepreneurship were set up within Bulgarian universities. It has been more than six years since the establishment of an advisory network by the Bulgarian Industrial Association and the university centres in Rousse, Svishtov, Varna and Sofia. These are the Training and Business Advice Centre (the University of Rousse), the International SME Centre (D. Tsenov Business Academy, Svishtov), the University SME Advisory Centre (the Technical University of Sofia), the International Centre for Small and Medium-Sized Businesses (University of Economics, Varna) and the Consultancy Centre (Institute for Postgraduate Training to the University of National and World Economy, Sofia).

b) Czech Republic

The Czech government has recently embarked on official programmes for innovation. For example, in April 2003 it decided to formulate an innovation policy by the end of 2003, which is nearing completion by the time this paper is being prepared. The draft of the National Innovation Strategy was prepared and drafts of the individual strategies for the individual sectors were worked out and approved by the individual ministries. The individual strategies that will be part of the National Innovation Strategy are now in the process of preparation.

In addition, the Ministry of Trade and Industry is preparing a new Operational Programme for Industry and Enterprises. This programme, which is currently in the process of preparation, will be focused mainly on SMEs and is designed to support both innovation infrastructure, such as science and technology parks and business incubators, and development projects of individual enterprises.

In April 2003, the Czech Government approved a new set of R&D programmes. The POKROK (CZ 15) programme will focus on fostering cooperation with foreign research organisations and firms and hence have the Czech Republic become part of the European Research Area (ERA). TANDEM (CZ 16) is a programme designed to cover needs of the research oriented towards industry sector. IMPULS (CZ 17) is a programme for industrial research, and mainly its development. All programmes were launched this year but the actual support of the individual projects will not start until 2004. The programmes are expected to be concluded in 2010 (2009 for the POKROK (CZ 15) programme).

Overall, the government's innovation policy appears mainly to focus on industry and Small and Medium-sized Enterprises (SME). The policy for support of SMEs till 2004 and the principles of the support programmes for small and medium-sized enterprises in the period 2001 to 2004 were jointly adopted by government resolution no. 1257 on 11 December 2000.

The 'Policy for SME until the year 2004' and the 'Support Programmes for SMEs for the period between 2001 and 2004' were approved as part of a document by the government resolution no. 1257/2000. The support programmes implemented the above-mentioned policy in practical terms. The long-term goal of the policy is to facilitate a functioning SME sector. Moreover, the policy should contribute to the following factors that are basic preconditions for the economic development of society:

- the performance of the national economy
- the sector's technical standard
- the sector's competitive strength as well as the reduction of unemployment and of regional, social and economic differentiation.

The medium term goals of the policy focus particularly on the improvement of the business environment for the development of SMEs in the Czech Republic. They include access of SMEs to:

- capital resources,
- information and information technologies
- education, advisory and consulting services
- technical standardisation and certification
- results of research and development.

They also focus on the reduction of the administrative burden in the process of establishment and development of small and medium-sized enterprises and on the creation of a favourable taxation environment.

c) Estonia

Estonia has the reputation of being a developed and economically open EU accession country with a stable currency, a balanced state budget and a liberal trade policy. The state's priorities in the last few years have been to join the European Union and NATO. Therefore many changes in legislation and the institutional system have been implemented.

In order to ensure economic growth, the government has paid rather a lot of attention to research, development and innovation policy issues – creating a favourable business climate and developing an integrated support system for business in general, and for innovative entrepreneurship. The favourable economic environment received an extra push with the Income Tax Act (started in 1 January 2000), which made all reinvested profit tax free. Statistics have shown a positive effect on the investment activity of companies, including investments directed towards R&D and innovation. During the last three years, several important documents have been developed and approved by the government and/or parliament. Most important are the research and development strategy 'Knowledge-based Estonia' (6 December 2001), and the small and medium-sized enterprises development strategy 'Enterprising Estonia' (15 January 2002). The latter defines the institutional structure and principles for state support of SME development from 2002-2006. The R&D strategy also set objectives until 2006 and government/parliament approval ensured the commitment to finance the planned strategy.

The R&D strategy measures are also included in the Estonian Single Programming Document (concrete action plan – September 2003) for applying for additional finances from EU pre-accession and structural funds. The R&D strategy foresees the creation and implementation of technology programmes in key areas – user-friendly information in the fields of technology, biomedicine and materials technology. The preparations for the biotechnology programme and advance planning activities in information technology and material technology began in 2003.

Several surveys and evaluations were implemented, because there were no good data available about the innovation activities of Estonian enterprises. The European Community Innovation Survey III was held in Estonia in 2001; results from the end of 2002 gave a better basis for policy development. On the institutional side, the whole state innovation system was reformed during the last 3-4 years. In 2000, the new enterprise support foundation 'Enterprise Estonia' started work. It is an umbrella organisation of five former foundations now redesigned as the Estonian Technology Agency, Estonian Export Agency, Estonian Investment Agency, Regional Development Agency and Tourism Agency.

New schemes to support enterprises were developed and provided through these agencies. The export and loan guarantee foundation KredEx also started work in 2000. Another big institutional reform influencing innovation policy in Estonia comes from the Government Research and Development Council (RDC), which started to act with the new structure from 2002. RDC is the main counselling body for the government on research, development and innovation issues. It has two sub-commissions: one in the Ministry of Education and Research responsible for research-related issues, and one in the Ministry of Economic Affairs and Communication responsible for innovation policy development. Recently, there have been discussions to widen the scope of ministries involved, including the Ministry of Social Affairs and the Ministry of Finance. In a similar way to the RDC sub-commissions, the two ministries divide their responsibilities to develop the policy and measures to support research and innovation. The grants for basic research are provided by the Estonian Science Foundation. The grants and loans for applied research and technology development are provided through the Estonian Technology Agency (from October 2003 onward the responsibilities of the Technology Agency will be handed over to Enterprise Estonia's new structural units). Technological development is also supported through support programmes like SPINNO (ES 17/launched in 2001), and the Competence Centres Programme (launched in 2003).

d) Hungary

Hungary has an internationally recognised, high-level research tradition both at university and academic level. Moreover, it has a good track record in natural sciences, engineering and medical sciences. Several international companies with R&D activities have chosen to locate to Hungary. Hence, from a private industry point of view, research is well integrated in international R&D networks. On a public level this is also confirmed by Hungary's participation in the EU Sixth Framework Programme. Public expenditure on R&D remains low and is predominantly state-funded. R&D infrastructure ispoorly developed and research staffs are of an ageing generation. In the highly competitive international markets the innovation activity of the Hungarian corporate sector is low. Above all the link between the R&D sector and businesses is weak: spin-off activity is minimal. This is also compounded by the complexity of the present innovation system, which seemingly does not allow for transparency.

It has been a common demand of individuals involved in Hungarian R&D that governmental tasks and responsibilities should be clarified, and that there should be access to a stable source of financing. A first good step forward was the government's programme issued in May 2002, which declared Research and Development and Innovation to be a priority sector. The creation of innovation-friendly regulation, making Hungary an attractive R&D location for investment, strengthening the protection of intellectual property and financing the innovation sources of SMEs are the most important areas of the programme's innovation policy. Moreover, after years of collecting data and surveys, a new national system of governance in the field of R&D policy has been developed. The new *Innovation Act* was submitted to parliament during September 2003. The new Innovation Act will remedy existing weaknesses in the public innovation system, as outlined above, and will hopefully become effective in 2004. It will provide the presently absent innovation institutional system, clarify innovation-related responsibilities and tasks on both governmental and local level, and it will also create the necessary legal environment, guaranteeing the stability of R&D financing.

The new Innovation Act also envisages an improved regional dimension, with the aim of strengthening cooperation between the central governments, local governments, regional institutions, and business partners. With the closer economic integration of EU countries there will be an increasing demand for a faster dissemination of results in the field of R&D. According to major Hungarian programmes (such as the National Development Plan and the Economic Competitiveness Operational Programme) there will be rapid development of the high-technology sectors, and the weight of knowledge-intensive sectors will also increase.

According to the Economic Competitiveness Operational Programme (ECOP), the Hungarian R&D system will be facing some major threats in the next few years, which could have consequences on the results produced by the present system. In the past two years the economic development of Hungary's main markets (EU, and specifically Germany) slowed down and so did the increase of income to the state budget, which potentially could have effects for innovation financing. Moreover, the brain-drain is still one of the main problems of the Hungarian R&D system. There is also an increasing gap in IT use between various sections of society and especially in rural regions, which are still falling behind in using IT-related technology. In 1999, after several years of declining financial sources and a decreasing number of R&D staff and units, the Hungarian Government focused on a gradual increase in expenditure for research and development, both by direct budgetary resources and indirect means, such as economic policies encouraging business innovation activities.

According to the Ministry of Education the expected impact of the Innovation Act is widespread. First, it creates a more effective and transparent structure in the fields of R&D and innovation, which has been lacking in Hungary so far. Moreover, there will be a much better coordination between the key elements of the Hungarian innovation system and improved efficiency (Havas, 2002). Through the result of the Act the different types of innovation clusters will be strengthened and cooperate with each other in the regions, and there will be a more effective protection of IPR. The establishment of the Fund will eliminate the contradictions between the annual budget points of view and exchequer management versus financial features of the long-term R&D projects *(financial dimension).* Moreover, the Act will contribute to the creation of new jobs and mobility of researchers.

The bill and the new Innovation Act state that the National Research and Technology Office (NRTO) will be at the centre of the new Hungarian R&D system. This office will elaborate governmental strategy in the field of innovation as well as shape all the means and tools for R&D and innovation policy at governmental level. The NRTO will coordinate the activities of the Research and Technology Innovation Fund and supervise the Agency for Research Fund Management and Research Exploitation. The task of the NRTO is make fundamental changes to the innovation attitudes of Hungarian society. Last but not at least, the NRTO keeps contact with other organisations in the governmental and civil spheres that are involved in innovation. Moreover, it will represent Hungarian interests at the international level such as the EU, and at all S&T affairs. From January 2004, the Research and Technology Innovation Fund will be a separate state fund under the control of the Board of the Research and Technology Innovation Fund. It will collect payments from enterprises based on a legal framework. Its main task will be to support the governmental R&D budget (as the legal successor of the current Technological Development Programme and the National Research and Development Programmes expenditures) through voluntary public interest payments of enterprises and private individuals.

The fund will provide financial assistance for industrial research and technological development activities. It will also promote the creation of R&D and knowledge-intensive jobs as well as other specifically innovation-related items. Additional tasks of the fund include the support of R&D activities of enterprises and supporting the supplementary R&D service sector (technology transfer, technology broker, PR, bridge-building organisations etc.).

e) Latvia

The Republic of Latvia is also on the path of transition towards a market economy. Relatively stable conditions (low inflation, minimal government debt, stable national currency, and positive balance of payments) have been created in order to continue the advancement of Latvia´s economic development. Privatisation of government-owned enterprises is nearing completion; growing economic activity and growth of exports resulted in a total GDP increase of 16.2% between 1995-1998. In 1999 GDP growth rates in Latvia, similarly to other countries, were influenced by the global financial and Russian economic crisis. GDP growth in 1999 as a whole amounted only to 0.1%. As the economy recovers, GDP growth is forecast to reach 3% in 2000. Latvia's accession to the European Union is one key priority for political and economic development.

During the last decade, however, Latvia's policy was mainly focused on transitional aspects, such as the privatisation of enterprises, the establishment of favourable economic conditions and attracting foreign investments. Relatively little attention has been given to policy making in the area of innovation and research and development.

Latvia's accession to the European Union is a key condition for economic development. During the last decade, Latvia's economic policy has increasingly transformed its focus from such transitional aspects as the privatisation of enterprises, the establishment of favourable economic conditions and attraction of foreign investments, to developing future vision, including innovation activities. However, as a result of the unstable and ever-changing political stage, efforts of the previous governments have not always been consistently carried forward.

The October 2002 parliamentary elections in Latvia brought on major changes in the composition of the government, revealing public dissatisfaction with the rate of development achieved by the last parliament. One of the strongest and oldest parties in Latvia, 'Latvijas cels', did not win a single seat in parliament, while the newly formed right-wing 'Jaunais Laiks' won a majority, after a strongly nationalistic election campaign. As a result, the current composition of Latvia's parliament can be characterised as noticeably youthful (several ministers are just over 30 years old) and with more women in high posts than before.

As a result of major changes in the leading ministerial bodies established by previous governments, innovation policy has also experienced some developments, observed mostly in the government's priority to cooperate in policy-making and the implementation process.

Since 2001, the most important innovation-related document for Latvia has been the National Concept on Innovation, which detects the problems in current institutional structures and estimates the consequences of not having an innovation policy in place. The National Concept outlines the benefits of innovation policy in dealing with current economic problems, improvements for the budgetary mechanism, the advantages of setting up an institutional system for development and coordination of innovation-related activities, and the development of a National Innovation System.

In February 2003, the committee of cabinet ministers approved the National Innovation Programme for 2003-2006, which was developed in 2002 on the basis of the National Innovation Concept. At this stage it is the most up-to-date and focused document in this sphere, having been prepared by the Ministry of Economics. Its goal is to implement targeted action towards developing a political, economic, social and cultural environment conducive to innovative development and growth in demand for knowledge-based services and economic activities. The programme includes a strategy for carrying out tasks from 2003-2006. Furthermore, it includes a mission statement, objectives, evaluation scheme and a list of actions and resources necessary for implementing the programme.

Next to the above-mentioned documents, the most noteworthy activities in the given time period are the reforms in education, further training and taxation, and the strengthening of research carried out by companies, resulting in the following documents: the order On Sustainable Development Baselines for Latvia, the Law on Innovative Education, the Latvian National Employment Plan for 2003, the Law on Innovative Education, the order On Baselines For The Development of Taxation and Revenue System and the Law on Customs Tax. All of these documents are discussed in greater detail throughout the report and in the attached measure sheets. Furthermore, several schemes and measures, including the NDP, will most likely be substituted by the new Development Plan that is currently being drafted by the Ministry of Economics before the end of the year 2003.

Important progress has been made in the following areas:

- raising public awareness and involving those concerned (National Innovation Programme 2003-2006)
- protection of intellectual and industrial property (Order on Conceptual Changes in Intellectual Property Protection Laws)
- innovation financing (Law on Customs Tax), and
- strengthening the ability of companies, particularly SMEs to absorb technologies and know-how (Latvian National Employment Plan for 2003).

Some measures have affected innovative activity without its being a primary goal. Such secondary advancements have been made in the areas of mobility for students, research workers and teachers, public authorities in support of innovation policy-makers, promotion of clustering and cooperation for innovation, administrative simplification, amelioration of legal and regulator environments, strategic vision of research and innovation, start-up of technology based companies, and intensified cooperation between research centres, universities and companies.

Within this time period, changes have at least had an effect on fostering innovative organisational and management practices in enterprises and public authorities, as well as supporting innovation policy-makers. Most of the changes in legal acts in Latvia at this time are geared toward the harmonisation of laws with EU standards and requirements. EU accession possibly remains the leading motivation for implementing changes and advancing innovation policy.

f) Lithuania

After Lithuania regained its independence in 1990, the period of transition from planned economy to market economy began. The government policy was focused, during the early transition years, on introducing market economy mechanisms, including privatisation and restructuring of state-owned enterprises. However, little attention was paid to developing an innovation policy. From the very beginning EU integration was recognised as one of the main priorities. After Lithuania became an associated EU member (in 1995), Lithuanian policy started to follow EU principles. The harmonisation of Lithuanian legal and economic environment became more and more intensive. Among other things it resulted in Lithuania joining the EU Fifth Framework Programme in 1998. From then onwards the importance of innovation in economic development was recognised. In 2000, the Government approved the first policy document, specifically oriented to innovation development – the Programme for Innovations in Business. After that the attention on innovation in Lithuanian policy grew more and more intensive. The importance of innovations for state development was recognised in the main policy document defining state development – the Long-term Development Strategy of the State (12 November 2002).

Currently, the government is taking further concrete actions in the sphere of innovation development. One of the recent policy documents is the Renewed Programme on Innovations in Business, approved on 15 July 2003. Essentially the main government focus is on creating an innovation-friendly environment for innovation development and the implementation of this in SMEs. One of the main priorities is the stimulation of cluster formation; the government lays much emphasis on creating the necessary cluster infrastructure, and fostering relations and cooperation between science and industry. However, the government is still struggling to create an effective financing system, designed for innovation start-ups. There are few taxation incentives specifically designed to support innovation development or to encourage science research. Overall, Lithuania has made notable progress in innovation development; however, some underdeveloped spheres remain, which should receive future attention.

During the past few years, Lithuania's governmental bodies took several actions, which impacted innovation policy in Lithuania. On 1 October 2002, the Government established the Science and Technology Commissions under the government of Lithuania. The commission is an internal advisory body of the government, which considers and submits to the government proposals for the creation and implementation of policies, strategies for the development of applied sciences, technology and innovations and stimulating interaction between science, industry and business. This governmental body should introduce more coordination for policy creation in the field of technology and innovation development; however, it is not yet a central institution that would be in a position to cover all innovation aspects.

On 12 November 2002 the parliament approved the Long Term Strategy of the Development of the State. This is a basic policy document for the state, and it encompasses all spheres of life. Therefore the attitude of the state to innovation development and main policy principles may be evaluated by analysing how much emphasis is put on innovation development in this document. It is a very general document, from which other concrete strategies, plans and measures are to be derived. The document foresees the main guidelines to be followed in developing the spheres of industry, education, science, governance and administration for the period until the year 2015. It contains an analysis of the existing situation, an evaluation of the strengths, weaknesses, opportunities and threats to development. The vision of the state in the spheres of science and education, economy, social policy, national defence, finance, culture and other spheres of life is also laid down in this document. The non-existence of any system of cooperation between science and industry, which would work according to modern models of innovation implementation, a low level of computerisation, limited involvement of high technology in industry and the inefficient infrastructure of business services are indicated as the main weaknesses. The document emphasises the following main strengths and potentials in Lithuania: the network of science and education institutions is sufficiently developed, with the result that Lithuanian science potential is rather high, and the rapid growth of the ITT sector.

On the 3rd December 2002, all governing parties represented in parliament gave their approval to the National Agreement Striving for Economic and Social Progress. This document is a general agreement between the policy makers establishing major policy guidelines as a priority for their activities. There are six actions defined, which should be undertaken in the next four years. Formation of clustering, economic policy, formation of institutions which should create high-tech policies, and the creation of possibilities to develop high-tech industry are among them. The issuing of this document illustrates the importance of recognising an innovation-oriented policy for Latvian policy makers. The main architects of the policy take innovation development as a priority that is guiding their political activities. On 24 June 2003, the Law on Investment was amended. The amendments introduced new ways and new objects of state support, which should foster the development of clusters and innovations. On 11 July 2003, the Government approved a draft of the statutes of Science and Technology parks. This is a document which should serve as the basis for the functioning of science and technology parks and so will help with the establishment and operation of such parks.

On 15 July 2003, the Government adopted the Programme on Innovations in Business (LT 16), which supersedes the analogous document (LT 08), adopted in 2000. The new programme is a policy document, specifically oriented to innovation development in the business sector. It describes the existing situation of the innovation system.

The main problems addressed are the fragmentation of innovation-related information and the unsystematic and unlinked spread of information sources; the insufficient cooperation between science and industry; the inexperience of science institutions in working with business entities and insufficient financing of innovations. The document also lists concrete measures to be implemented during 2003-2006 that should foster innovation development in the country; therefore it is a strategic measure at the same time. It addresses the fields of science and industry cooperation, improvement of qualification and education of society, the financing of innovation activities, reform of innovation administration, etc.

On 18 July 2003, the Government approved the Concept of Developing Science and Technology Parks. This policy document is specifically oriented to the development of science and technology parks in the country. It lays down main priorities, goals and objectives, which should be observed and achieved in order to encourage the creation and development of science and technology parks and intensify the cooperation between science and business.

g) Poland

In the early 1990s the legal foundations for a market economy were built. However, privatisation and the inflow of foreign capital are not universal remedies. The economy has some weaker spots in the mounting state budget deficit, in the low competitiveness of domestic industry, in the high unemployment rate and in the overall R&D inefficiency. During the hard negotiations on European Union accession, the government realised that it must initiate a more active economic policy, similar to that in the EU.

One of the first signs of this new approach was the approval in 1999 by the Council of Ministers of the strategic document *Zalozenia polityki innowacyjnej panstwa do 2002 roku [Guidelines for the state innovation policy until 2002]*, prepared by the former Committee of Scientific Research. The document contains a set of directions with associated tasks which are important for Poland's innovation policy. The tasks to be implemented are described precisely, with suggested deadlines and institutions responsible for their fulfilment. However, no information about financing sources was provided. Also in 1999, the government approved the document entitled *Kierunki dzialan Rzadu wobec malych i srednich przedsiebiorstw do roku 2002 [Government policy guidelines for Small and Medium-sized Enterprises until 2002]*. This is a programme which defines the economic measures and actions to be undertaken in order to establish favourable conditions for the development of the SME sector.

Creating these conditions should increase, in turn, the ability of SMEs to invest more in new technologies and to introduce more innovations in order to compete successfully on international markets.

In 2000, the Ministry of Economy set up a far-reaching programme entitled *Zwiekszenie innowacyjnosci gospodarki w Polsce do 2006 [Increasing the innovativeness of the Polish economy until the year 2006]*. The programme was established to ensure that, during the pre-accession period, the government, as well as non-governmental institutions, will not focus merely on remedying the effects of past neglect but also on developing more specific commitments, which will lead to the narrowing of the gap with other EU economies. Within the programme, the Ministry has identified four priority areas:

1. Creation of mechanisms and structures conducive to innovation
2. Stimulating innovative attitudes
3. Increasing the efficiency of implementing modern technologies in the economy
4. Making consumption standards and production models more favourable to long-term development.

For these priority areas, a number of improvement-focused sub-programmes have been proposed, together with suggested action plans and institutions responsible for their implementation, as well as for the requested budget. The year 2002 was a period of intensive work on a set of new strategies which are to be implemented after the EU accession. The most important are:

- Strategiczny Program Operacyjny – Wzrost konkurencyjności gospodarki (Strategic Operational Programme – Increasing the Competitiveness of the Economy)
- Zintegrowany Program Operacyjny Rozwoju Regionalnego (Integrated Operational Programme of Regional Development)

The above strategic documents – after their adoption by Poland's Committee for European Integration – are currently being reviewed by the European Commission. Looking back over the period 2000-2003, it may be noted that strategic documents have been growing in numbers. But the level of their implementation was extremely low and resulted only in a few measures aimed at improving the innovativeness of the economy. Recently only two measures are worth mentioning:

- PL 13 Financial support for new investments – with the aim of supporting large and medium-sized companies with innovative projects. The measure is administrated directly by the Ministry of Economy, Labour and Social Policy, with the state budget as the source of financing.
- PL 14 Grants for entrepreneurship development – intended as support for the SME sector. The measure is administered by the Polish Agency for Entrepreneurship Development (PARP), and it is financed both from the state budget and EU aid funds.

After the Copenhagen summit of December 2002, the deadline for Poland's expected accession to the European Union (1 May 2004) has been approaching very quickly. Unfortunately, the distance separating Poland from the EU in various areas will not be decreasing with the same speed. It is worth analysing how the situation is regarding some matters related directly or indirectly to competitiveness and innovativeness:

a) GDP per capita: In 2000, Poland achieved the level of US$9,685 (around €7,980 at Purchasing Power Parity). The same year the average EU GDP per capita was US$24,400 (around €20,120). More important may be that in the last couple of years Poland has been growing slower than the rest of the 10 accession countries. In 2001, GDP growth was only 1%, and 1.3% in 2002, and the distance from faster-growing accession countries will be more and more difficult to make up.

b) Foreign direct investment (FDI): For Poland, FDI is a very important factor of GDP growth and the source of new technologies. From the beginnings of economic transformation in 1989 Poland had the highest inflow of FDI in absolute terms in the region. In 2002, according to the PAIZ (State Agency for Foreign Investments), accumulated FDI amounted to €65.4 billion but, in the same year, Poland attracted only €6.5 billion of FDI which was a large drop compared with 2000 (€11.5 billion) and less than the capital invested in the Czech Republic in year 2002 (€9.1 billion). Moreover, the PAIZ statistics seemed to differ from reality, as they quoted the values declared by investors, and even the National Bank of Poland (NBP) was quoting lower values based on effective transfers of capital. Also the international UNCTAD statistics do not paint a better picture of Poland. UNCTAD quoted €165 per capita in 2001 (inflow of FDI) and €1,180 per capita (accumulated FDI). The same per capita indicators for the Czech Republic are: €3,536 (inflow FDI) and €2,945 (accumulated FDI). For Hungary, it was €268 (inflow FDI) and €25,700 (accumulated FDI). According to economic analysts, Poland would need about €10.6 billion per year to reach a GDP growth of 5-6% per year.

c) Exports: In 2001, Poland exported goods worth €1,039 per capita. The same year, the average exports from the EU were €6,632 per capita. Again, other Central European countries were much more export-oriented than Poland: Czech Republic exports were worth €3,613 and Hungary's €3,325 per capita.

d) Expenditures on R&D as percentage of GDP. In Poland's case, this was 0.65% of GDP in 2001, and most of these expenditures come from the state budget (0.45%), as the state finances the whole sector of research institutes, laboratories and university research carried out in universities. In comparison, currently the EU average is 1.95% of GDP and the record result of 3.8% belongs to Sweden. This area will pose a problem for Poland as the Lisbon Strategy assumes 3% of GDP in 2010 of which 1/3 must be financed by the state and 2/3 by the private sector.

e) Innovative companies. According to the official GUS (Central Statistical Office) statistics for 2001, in Poland the proportion of innovative companies in the total number of registered companies (companies with more than 10 employees) is 16.9%. In the EU, 51% of companies are considered innovative.

f) Innovativeness: In 2000, the average indicator of innovativeness was 0.6 for Poland and 2.6 for the EU. Poland is not a member of the European Patent Organisation (EPO) and Poland's Patent Office works ineffectively and is in structural crisis. Poland is also lagging in information technology, with 1.2 Internet hosts per 100 people. However, looking at the structure of expenditures planned in the 2003 budget, one notices that the subsidising of the computerisation of the country, as well as of the science sector, is only symbolic (an allocation of €27.5 million on computerisation and €392.4 million on science). Instead, state assistance is earmarked for unprofitable coal mining (an allocation of €236 million), while a considerable part of the resources are going to education (an allocation of €1,538 million), and will be spent on swollen public administration (€1,455 million).

This all means that the government's economic strategy in fact pushes Poland to the margin of modern Europe, as without gigantic investments in science in general, and in tele-information technologies in particular, or the engines of the knowledge-based economy, Poland has no chance to leave its current position of an outsider. The distance that the Central East European region has from the more advanced EU countries is now being made up in Estonia, Hungary and the Czech Republic more rapidly than in Poland.

h) Romania

The overall importance of innovation policy in Romania has increased during recent years. After its launch in 1996, the National Programme 'HORIZON 2000' became the principal instrument for the orientation of research activities. The programme was elaborated on the basis of input from actors in the research system and represents the sum of the existing and potential capacities of the research institutes. This programme proved to be insufficient in terms of the areas and economic domains that were financially supported. For this reason, in 1997 a new system of Research, Development and Innovation (RDI) financing was put in place – the National Plan for RDI (regulated by GD 48/1998, and GD 562/1999). In order to avoid negative consequences that may appear during the transition from one financing mechanism to the other, Law 191/1999 prolonged the operability of the programme 'Horizon 2000' until the year 2002.

This allowed financing for programmes started under the old financing scheme to continue without interruption. As a result of the updating of the National Plan for RDI for the period 1999-2005 though GD 556/ 7 June 2001, the number of RDI programmes, covering a wide range of economic domains, has been extended to 14.

Progress has been made as regards fostering an innovation culture, and establishing an innovation framework. However, further measures are needed to increase application of research results by businesses. For that reason, one of the priorities of governmental policy in this domain is to develop the interface between academia and the business environment in order to secure better conditions for technology transfer.

The Romanian government has set the development of the ITC sector as a strategic priority for the national economy, aimed at transforming Romania into a high-tech area that would be used for outsourcing purposes in the Information Technology domain. Growth in this area is about 15% per year – much higher than other sectors of the economy. The state also intends to support focused objectives associated with a particular industry, such as IT. The main target industries, which the government emphasises, are the ones belonging to the high-tech sector (i.e. software development, networking, hardware components manufacturing etc.). There are plans to have a software park in every Romanian city that has a university centre.

Building Romania's competitiveness requires the intensification of a complex set of common activities at the level of the firm and the industry cluster, reinforced by policy and institutional action at the level of national and local government. Delayed engagement in higher technology and adding higher value to business activities are proving detrimental to Romania's competitiveness. Among the major impediments to the competitiveness of the Information Technology and Communications sector in Romania are:

- Lack of capital in the country to support the expansion of existing businesses and creation of start-up businesses
- Lack of a sizeable domestic market that could stimulate the sector's development or attract IT multinationals into forming cooperation relationships with local partners to serve such a market
- Limited participation of foreign investors who could introduce managerial know-how and technology to the industry
- Lack of entrepreneurship, marketing and management skills that could bring more international exposure and experience of domestic firms
- Limited R&D base, and insufficient investment in R&D for new technology and software
- Lack of private sector consultation and policy formulation to support the industry as well as R&D activities.

(Centre for International Private Enterprise, 2001)

The main measures that stimulate the cooperation between research and industry are the programmes run under the National Plan for Research, Development and Innovation (RDI). It is worth mentioning that, during 2001, approximately 85% of the projects selected for financing included a partnership between Research & Development units and businesses, with co-financing of the latter (the co-financing weighting being around 34%). In 2001, the total number of projects submitted under the National Plan for RDI was 4,618, of which 1,302 received financing. By comparison, in 1999 these indicators were 2,789 and 504 respectively.

i) Slovakia

Immediately after the political turnover in November 1989, many changes were carried out inside research institutions: new management, more democratic bodies in the control system, restructured research programs, etc. At the level of the R&D system on the whole, the point of departure for the transformation was the scenario for economic reform from a strongly centralised planned economy to a market one. According to this general scenario a new strategy of scientific development had to be implemented with the aim of abolishing obstacles in the sphere of new knowledge production and in the innovation process, in order to facilitate free science creation as well as the R&D system compatible to a market economy. This new strategy became a part and parcel of the transformation process as a whole. In October 1990, still within the framework of the Czech and Slovak Republic, the federal government adopted the principles of state policy in the sphere of science. It proclaimed a willingness to guarantee freedom of scientific and technological activities, to respect the principle of autonomy in this field and to accept the concept of financial support plurality (Šarmír, Zajac and Dušan, 1999).

Many other restructuring measures, above all in the field of funding, were approved at the state level, both in the framework of the Federal State and of the independent Slovak Republic after 1993. They were mostly partial in character, and it is necessary to emphasise that up to this day no comprehensive conception of the R&D transformation has found general acceptance or been officially adopted. The situation has evolved rather spontaneously. The research institutions made use of their newly acquired autonomy to adapt themselves to the rapidly changing situation. In particular, they had to react to the strong impact of the state budget policy that has led to a serious scarcity of resources. They were required to react to new types of financing as well as a new possibilities of international contacts and cooperation.

Such a development caused a rapid decrease in the R&D research potential in the social sciences. There was a 50% reduction in 1996 as compared with 1989. Important was the fact that the changes introduced after 1989 aimed at granting greater autonomy to educational and research institutions and sheltering them from direct political interference. A second important change was to be seen in opportunities of developing closer contacts and cooperation with major scientific institutions in developed countries.

Table 1 summarises the trends identified in the survey.

Table 1: General status of innovation in the surveyed countries

	Country	Status
A.	Bulgaria	Innovation at a basic stage
B.	Czech Republic	Innovation at a basic stage but policy provisions in place
C.	Estonia	Innovation at a basic stage with policy provisions and tax incentives in place
D.	Hungary	Innovation at a basic stage but policy provisions in place
E.	Latvia	Innovation at a basic stage but policy provisions in place
F.	Lithuania	Innovation at a basic stage with policy provisions and tax incentives in place
G.	Poland	Innovation at a basic stage with policy provisions and tax incentives are in place
H.	Romania	Innovation at a basic stage but policy provisions in place
I.	Slovakia	Innovation at a basic stage but policy provisions in place

It can be seen that many different stages of evolution are seen, but since this is a dynamic process, things are likely to change in the new future.

A question of funding?

The previous section of this paper illustrates the fact that Europe needs to energise the transfer from academia to industry and professionalise it. This is in line with findings from many countries in the region as outlined by Leal Filho (2002, 2004). While R&D is among the most important sources of wealth in the EU, over the last ten years many European R&D jobs have been lost to the US. A good example of this is the pharmaceutical industry. In 1992, six out of the top ten medicines were developed in the EU, compared to four in the US. By 2000, the US was radically in the lead, having doubled that figure, which shows the benefits of investing in promising, technologically demanding sectors.

The emphasis for innovation in Europe has so far been limited to the low risk, low added value environment, generating few transformational innovations and hence reducing the European competitive position on the global scene. If Europe wants to regain its competitive edge, it must aim to make investment more attractive. European companies should also shift external costs into internal investments and move from incremental innovation to transformational innovation.

The 3rd European Business Summit (EBS), held on 11-12 March 2004 in Brussels, acknowledged that in order to achieve this companies must link their innovation efforts to the creation of economic value. They must innovate within a clear strategic perspective and provide substantial added value to their customers. Last but not least, they must be driven by clear leadership with experience in transforming 'findings' into 'value'. Furthermore, the right partnership must be built between academia and industry, and both must share the long-term vision, comprising both risk and gain. Publicly funded research organisations also need good governance at senior management level, participants agreed. Management should be more accountable, with more business expertise, and should be in direct contact with the demand side – industry. In parallel, public authorities should reduce taxes related to R&D investment, business investment, personal taxes for scientists, and encourage a move from project-based funding to long-term structural incentives and knowledge creation.

In a further reference to issue of financing, the 3rd EBS called for support for the so-called "boundary conditions". These are described as taxes, financial incentives, labour market, education, environmental regulations, permits and technology transfer from university research. Without stability and reliability in boundary conditions, it is acknowledged to be difficult to plan an innovation strategy running over a number of years. Innovation experts also call for government policy-makers to improve the fiscal treatment of R&D and the fiscal incentives for carrying out R&D; to improve existing R&D support and lessen the bureaucracy associated with it; reduce the regulatory hurdles such as obtaining permits and acquiring patents; and stimulate a greater application and industry orientation of public research at universities.

Conclusions

This paper has given an updated overview of the state of play on innovation in various Central and Eastern European countries. From the state of affairs it describes, some future tendencies can be identified. These are as follows:

a) the promotion of clustering and cooperation for innovation

There are noticeable development in this area, as organisations and companies are beginning to recognise the advantages of cooperation and clustering of enterprises. In various countries, there has been an emergence of institutions to provide management services to SMEs, and help to establish partnerships between companies. One example of an institution doing this is LTP in Latvia, which aims to organise scientists and highly qualified professionals to support the realisation of new ideas. LTICA's principal objectives are:

1. To create permanent links among all the business support structures in Latvia;
2. to create an information network to support technology-oriented SMEs;
3. to promote the establishment of new business support structures;
4. to create contacts and participate at various different international and regional associations;
5. to cooperate with governmental and local authorities, science and research centres.

Such institutional set-ups are needed in other countries, so as to provide additional guidance to innovation.

b) the promotion of regional cooperation

A second trend seen is towards supra-national, i.e. transborder cooperation in the field of innovation. In this context, the Baltic Association of Science/Technology Parks and Innovation Centres (BASTIC) for example, which unites institutions in Latvia, Lithuania and Estonia in one organisation, may play a key role. The members of BASTIC are active participants in the ICECE (Innovation Centres in Eastern and Central Europe) Work Group linking representatives of national associations and innovative business support structures from Central and Eastern European countries.

c) the need to strengthen institutions

It is also necessary to strengthen the capabilities of R&D institutes as a prerequisite to conducting the "national R&D programmes" and increasing the number of R&D personnel in both public and private sectors. Again, an overriding emphasis is put on the "supply" side, while quite a few important players and elements of the innovation process are eclipsed.

d) the need to provide long-term funding

No matter at which angle one looks at innovation, be it in the field of education, infrastructure, industrial, SMEs, regional, health care or the environment, projects in this field require substantial investments and/or generous subsidies. The transition process, however, which has hit many Central European countries in a hard way, means that they are not always able to cope with significant investments. However, this is necessary if they want to be regional players, bearing in mind that success in innovation means economic growth and a better situation in dealing with social problems such as unemployment.

Finally, there is one final matter countries should be mindful of. This is the need to avoid the "good old" linear model of innovation, indicating the strength – as well as the way of thinking – of the "science" lobby. The systemic, complex nature of innovation, even the basic concept of demand for innovation, should be kept in mind at all times. Research, development and innovation should always go hand in hand and should be seen as such in national development strategies.

References

Centre for International Private Enterprise (2001) Romania – National Competitiveness Report. CIPE, Bucharest.

European Commission (1993) An integrated approach to European Innovation and Technology Diffusion Policy. A Maastricht Memorandum. European Commission, Directorate-General XIII, Brussels-Luxembourg, May 1993.

European Commission (1995) Green Paper on Innovation. European Commission, December 1995.

European Commission – Directorate-General Enterprise (2003) European Trend Chart on Innovation. European Commission, Brussels.

European Commission – Directorate-General Enterprise (2003a) European Trend Chart on Innovation. Country Report Bulgaria. European Commission, Brussels.

European Commission – Directorate-General Enterprise (2003b) European Trend Chart on Innovation. Country Report Czech Republic. European Commission, Brussels.

European Commission – Directorate-General Enterprise (2003c) European Trend Chart on Innovation. Country Report Estonia. European Commission, Brussels.

European Commission – Directorate-General Enterprise (2003d) European Trend Chart on Innovation. Country Report Hungary. European Commission, Brussels.

European Commission – Directorate-General Enterprise (2003e) European Trend Chart on Innovation. Country Report Latvia. European Commission, Brussels.

European Commission – Directorate-General Enterprise (2003f) European Trend Chart on Innovation. Country Report Poland. European Commission, Brussels.

European Commission – Directorate-General Enterprise (2003g) European Trend Chart on Innovation. Country Report Romania. European Commission, Brussels.

European Commission – Directorate-General Enterprise (2003h) European Trend Chart on Innovation. Country Report Slovakia. European Commission, Brussels.

Havas, A. (2002) Does Innovation Policy Matter in a Transition Country? The case of Hungary. Journal of International Relations and Development, vol. 5 (2002), No. 4, pp. 380-402

Högselius, P. (2003) Can East European Countries Innovate? Paper prepared for the DRUID Winter Conference, Aalborg, 16-18 January 2003, by Systems of Innovation Research Programme, Department of Technology and Social Change, Linköping University, Sweden.

Leal Filho, W. (ed.) (2002) Prospects of Integration and Development of R&D and the Innovation Potential of Black Sea Economic Cooperation Countries. IOS Press, Amsterdam

Leal Filho, W. (ed.) (2004) Supporting the Development of R&D and the Innovation Potential of Post-Socialist Countries. IOS Press, Amsterdam

Šarmír, E., Zajac, S., Dušan, A. (1999) Social sciences and the challenge of transition in Slovakia. In Faltan, L. (ed.). EUROPE: Expectations and Reality. The Challenge for the Social Sciences. Slovak Academy of Sciences, Institute for Sociology, Bratislava.

Viesturs, U., Draveniece, A. (1998) Innovation Problems in Economies in Transition: the case of Latvia. Eurolat, University of Latvia, Femirc-Latvia.

R&D Priorities in Innovation Policy and Financing in Former Socialist Countries
W. Leal Filho and P.S. Gramatikov (Eds.)
IOS Press, 2005

Chapter 2
Innovation in Europe – an Overview of Current Trends

LYNN M. MARTIN
University of Central England
Perry Barr G212
Birmingham
United Kingdom B42 2SU

Summary

This chapter reviews some of the trends around the current state of play in terms of the development of innovation and entrepreneurship in Europe during the process of enlargement. This includes a review of how trends established three years ago are still evident in EU/US comparisons. It also explores the way in which such comparisons are made and the emerging business models identified as important for innovation in the European Union.

Introduction: Innovation and the Need to become a Knowledge Economy

The current situation in the EU in terms of innovation relates to the drive for it to become a functioning knowledge economy across all member states. The knowledge economy is characterised as the route for national competitive advantage, with those unable to participate in this development being left behind. Basically, the "new, knowledge-based, post-industrial economy" has a number of common factors represented in the literature:

- Knowledge rather than capital is the new primary factor of production in the new economy with intangibles are primary assets such that service and not goods is the predominant value driver and such assets a brands, technology, intellectual property are increasingly important
- Linkages and networking are increasing fundamental to working practice; since current information communications technologies (ICTs) enable widespread networks of both people and machinery, enhanced by the digitisation of data allowing diffuse transfer, storage and processing of information. This results in expansion of the possibilities of coordinated action and collaboration within and outside the organisation, region, country and industry.
- Internationally, the economy is subject to rapid change as a result of the rapid pace of innovation and also the impact of new communications technology, characterised by a huge reduction in product life-cycles and by the increasing importance of virtual aspects such as virtual money, digital signatures, "virtual transactions dissolving the boundaries between the real and the imaginary". (Grant, 2000; Stansbridge, 2003)

Background

Innovation remains a priority for all Member States and for the European Commission. Throughout Europe, "hundreds of policy measures and support schemes aimed at innovation have been implemented or are under preparation. The diversity of these measures and schemes reflects the diversity of the framework conditions, cultural preferences and political priorities in the Member States". These followed the development of the first 'Action Plan for Innovation in Europe', launched by the European Commission in 1996, which was intended to provide for a common analytical and political framework for innovation policy in Europe for the first time (EIS, 2003 annex a). Over the last ten years, major structural changes in the economy shifted Europe's focus from traditional industries to gain comparative advantage via knowledge-based activities. This was accentuated by the pressure exerted by the forces and influences accompanying globalisation.

Here, for instance, competitive pressure was exerted upon manufacturing firms in high-cost locations, leading to a shift in production capacity to lower cost countries and to increasing use of technology for productivity. During this period, the development of ICTs (Information and Communication Technologies) opened up new markets, via personal computers, software and ICT-based services. ICT-based services have revolutionised production processes in many industries and led to growth of the service sector (Green paper). ICTs also changed the way business easy done, especially in the service sector. At the same time and as part of an ongoing process, within the European Union the Internal Market continues to develop, barriers and tariffs have been and will be removed. This, is predicted will encourage harmonisation and enable trading in a market of significant size and promise given that it now comprises 450 million people.

However, if Europe is to take advantage of these new markets and of these continuing new developments, it will need to be able to show consistent and increased strengths in terms of entrepreneurship and innovation, with adaptability to change seen as crucial for competitiveness. Ambitious targets have been set to achieve this. In identifying that 15 million new jobs will be needed by 2010, the European Council identified key needs to radically transform the economy. All this has led to a focus on innovation and entrepreneurship. The Green Paper (2003) for European Entrepreneurship for instance highlights new opportunities for entrepreneurial initiative, particularly in the area of services.

Where manufacturing industry survives and prospers it is expected to do so by developing added value via the impact of knowledge. Similarly, the increased complexity of production processes requires a variety of specialised inputs while transmission of knowledge across space has seen continued cost reduction, making inputs by external providers relatively cheaper. Here increased market opening allows firms in niche markets to operate on a more European or worldwide scale.

One group expected to benefit from this is the small firms sector, comprising around 98% of all European businesses and with key attributes which seem designed to allow for niche market exploitation, including flexibility and closeness to the marketplace. In addition such firms are expected to be able to gain most competitive advantage from the development of ICTs since they will be able to compete with larger organisations on equal terms for the first time (Forster, 2000; Poon and Swatman, 1997). One group which is seen as particularly important comprises the "high growth" SME, which is expected to play a significant role in job creation. (OECD, 2002)

In order to evaluate how well Europe has progressed with its ambitious agenda to empower enterprise and enable innovation, it is necessary to examine the range of evidence, most of it accumulated and defined by the European Union itself as part of an attempt to benchmark Europe against key global competitors

The perception of current innovation performance – EU perspectives

Current information as to the relative performance in 'innovation' of the European Union does not make comfortable reading. In assessing performance the EU uses a range of measures to benchmark member and to produce a resulting innovation scoreboard. The use of the word 'innovation' may also in itself be problematic given that it is associated most often with invention rather than commercialisation of a new product or service. However Drucker amongst many identifies innovation as the brining to the marketplace of a new product, process or service to enhance an existing product or to develop a new enterprise (Drucker, 2001). The element of commercialisation is less readily measurable despite the emphasis on this aspect in the definition.

Attempts to evaluate European innovation usually relate to numbers of patents, numbers of graduates in science and engineering, investment in public and private sector research plus numbers of academic refereed papers published. These are the terms and measures used by the European innovation scoreboard (EIS). Set up at the request of the European Council (Lisbon, 2000), the European Innovation Scoreboard focuses on *high-tech innovation* and provides indicators for tracking the progress of the European Union towards the Lisbon goal of becoming the most competitive and dynamic knowledge-based economy in the world within the next decade. The EIS indicators are divided into four categories, containing 17 main indicators and 3 additional indicators for the Candidate countries, all selected to summarize the main drivers and outputs of innovations (See appendix A for further details).

However, performance is relative – old member states are compared with each other and with new member states and the EU itself is compared with the USA and Japan. In part this is historic, since comparisons with the USA have been part of economic analysis since the 1920s. However it also relates to the position of the USA in terms of competitiveness and productivity. Although Finland has appeared as number 1 in competitiveness worldwide (Martin, 2002), the USA has retained its position as "top nation" in terms of economic performance despite recent recession, high rates of borrowing and a payments imbalance.

Since 9/11 the USA may have appeared to falter but some recent indications show resurgence, especially in terms of increasing employment and expansion in the small firms sector. The increased levels of defence spending in Research and Development may also have a significant effect on longer-term research and innovation performance, given previous research linking defence spending with research cluster and corridor development in California and the East Coast over the last fifty years.

In comparing the EU with the USA, eleven indicators are used but the 2003 EIS contains 19 main indicators, selected to summarize the main drivers and outputs of Innovation. These indicators are divided into four groups: Human resources for innovation (5 Indicators); the Creation of new knowledge (4 indicators); the Transmission and application of Knowledge (3 indicators); and Innovation finance, output and markets (7 indicators). Data comparability issues emerge across these indicators, acknowledged as part of the discussion in the appendices attached to the EIS.

Regional differences

In 2002 EIS was expanded to include a Regional Innovation Scoreboard or RIS, which was limited to a static comparison of those indicators for which regional data was available. By 2003 the RIS also included more region-specific innovation indicators and a dynamic comparison of the regions. This year's RIS uses regional data for 13 innovation indicators plus per capita GDP at the regional level for the EU member states. The regional innovation indicators listed provide good coverage of the innovation categories

- *Human resources* (4 indicators),
- *Knowledge creation* (4 indicators)
- *Transmission and application of knowledge* (4 indicators).
- *Innovation finance output and markets (*limited to only one indicator).

Due to data limitations, the regional indicators are better at identifying strong innovative regions than regions with future potential, or regions that require diffusion-oriented policies. Despite such caveats, analysing data regionally is valuable as it allows better review and evaluation of the effects and impacts of varying innovation policies which have been developed and implemented in addition to national and EU level policies, at the micro level, i.e., at the regional and even municipal level, at the regional and even municipal level.

Identifying regional differences can inform regional policymaking to make it more effective. Empowering high growth small firms has been the aim of government policy directed towards ensuring appropriate framework conditions, with a focus on the creation of efficient markets and the removal of obstacles to the creation, expansion, development and exit of firms. Nationally, policies affecting business start up and expansion have been reviewed with respect to administrative measures affecting start-up, financial and fiscal measures affecting venture capital supply and stock options, intellectual property regimes governing ownership of intangible assets, the flexibility of labour market policies and the aspects of education and training policies that affect the pool of skilled workers. (OECD, 2002)

Examples of differences in policy implementation are the recent regional emphasis placed by Sweden and the United Kingdom, which lags behind the established policies of Germany, Belgium, Spain, and Austria, with their long tradition in this area. Similarly differences may occur in the way funding is allocated and accessed. Regional development problems can be addressed via additional funding from the European Regional Development Fund, with innovation promotion seen as a key factor for ERDF programmes. Regional indicators can help inform these policies by identifying trends, evaluating impacts and providing a benchmark on a regional basis.

The second impact of regional activities in innovation relates to clusters and the impacts of clustering on SME competitiveness. In both industrialized and developing countries there is increasing awareness that isolation, rather than size, is the key obstacle, preventing SMEs boost their competitiveness. Groups of firms located in close proximity (*geographic clusters*) or in related business sectors have proved to be capable of rapid economic growth, sustainable leadership in export markets, significant employment generation and/or preservation of high-value-added jobs, sustained technological progress. Evidence from both developed and developing countries testifies to the unique opportunity that SME cluster development provides for reconciling the objective of economic development, environmental sustainability and social equity. (UNIDO, 2003;Boari, 2001)

However it would be wrong to represent clusters as a new phenomenon. The industrial revolution in the UK for instance often depended on geographical clustering of production for economy of scale (the cotton industry has been cited as an example of this phenomena).

However new effects are seen in the types of firms able to connect and be connected via the medium of ICTs. It may be suggested that the positive effects of clustering cited above can be found in the many dynamic clusters worldwide, as an outcome of the co-operative linkages both between local firms and among local firms and their business partners (such as suppliers of plant & machinery, producers of raw materials, testing laboratories, financial institutions, industrial technical and management consultancy organisations, training institutions and local government agencies).

The clustering of SMEs may not be so easy to identify in developing countries. Recent Government of India reports cite 400 SME cluster based on business sector but cooperation within clusters is hard to find (MSSI, 2004; SIDO, 2004). "Very low levels of trust, latent conflicts, and absence of suitable discussion foray are the unequivocal markers of an "under-performing" cluster" (UNIDO, 2004). Addressing the issue of under-performance is requires more than a quick fix solution.

Although technology may aid communication these also require soft factors such as sensitisation, trust building, conflict resolution, etc for any real cooperation to occur. Here government has a role to play in overcoming these barriers in the private sector since external assistance based on public-private sector partnership can bridge the gap in trust, and therefore greatly facilitate improvement in the performance of "underachieving" SME clusters.

As has been explained, the EU Innovation Scoreboard for 2003 explores the development of the EU/US gap in terms of these indicators, which represent those for which comparable data are available. As was the case in 2002, the US outperforms the EU across 10 out of 11 indicators. Only in terms of the number of science and engineering graduates being produced does the EU outperform the USA. Ominously perhaps, the EIS also suggests that if the current rates of change remain constant, this situation would remain at least until 2010 – and beyond.

Patent issues

In exploring patent applications, the EIS measure used reviews and places emphasis upon the number of patents applied for at the European Patent national (and regional) offices. This is particularly related to high technology applications which are measured as applications per million population. Using EUROSTAT it can be seen that high technology patent classes include:

1. Computer and Automated Business Equipment
2. Micro-organism, genetic engineering
3. Aviation
4. Communication
5. Semiconductors
6. Laser: H01S.

This indicator covers research carried out anywhere within a firm and not just within a formal R&D laboratory. The indicator also measures specialization of knowledge creation in fast-growing technologies. For some countries the absolute numbers of high-tech patent applications are so small, that the relative level of performance is both close to zero and highly unstable over time. This last category includes many of the former east European states, with Latvia, Lithuania, Czech Republic, Poland, Bulgaria and Romania all being less than 1. This might be anticipated from the context and historical perspective, and the trends in these countries will be interesting to evaluate.

However, despite the better performance of some of the EU countries, notably Norway, Netherlands, Denmark and Germany, the patenting figures indicate the need for all round improvement. Here there is a specific European weakness given that the USA is patenting more actively not only in the USA but also in Europe itself. This is true across different types of patents. High technology patents show this same trend – a further difficulty given that the high technology sector is expected to lead the economy into competitive advantage. This is characterised as the most important segment for innovation, given the predictions that only the development of a knowledge economy will lead to continued growth and prosperity in the 21st century.

In knowledge economy advantage is gained via the development of knowledge as products as services. The decline of traditional industries is to be addressed by the development of a service-based and high technology based economy The key to developing a knowledge economy is seen in the encouragement of innovation, the empowerment of knowledge creation and knowledge development processes, and the establishment of an entrepreneurial culture.

For 2010, the Union aims "to become the most competitive and dynamic knowledge-based economy in the world, capable of sustainable economic growth with more and better jobs and greater social cohesion". Knowledge has been described as the main differentiator between success and failure (Raeside and Walker, 2001, p 159), and vital for innovation and growth (Perez-Bustamante, 1999; Swan et al, 1999). Knowledge may be *scientifically* or *socially* constructed; the recognition and use of socially constructed knowledge is important for innovation (Nonaka and Takeuchi, 1995). Practical and experiential learning processes aid innovative knowledge construction while employee interaction is knowledge creation (Marshall and Reason, 1993).

Although. the future introduction and implementation of the EU patent may facilitate the patenting process in Europe; this alone may not be sufficient to overcome the underlying patenting weaknesses in many Member States. The EIS proposes a concerted EU effort to boost European patenting in Europe and, more importantly, in the US. Certainly without his sort of effort change to this situation is unlikely; trend analyses show that Europe is unlikely to catch up in patenting in the foreseeable future without new and active measures.

The investment in research and development compared between 2001-2003 shows mixed results. While private sector investment in R&D shows some weak signs of recovery, the public sector spend on R&D. has not. Comparing the US to the EU, financing and investment is still harder to obtain in the EU. While early stage venture capital resources shows slow improvement, the gap still remains "huge". Similarly with respect to human capital, there is again a large gap in the levels of tertiary education. Assessing this fully is difficult since comparable data does not exist bet went the US and the EU for key factors such as lifelong learning.

But is there any good news?

As has been identified above, the only area where the EU leads the US is in terms of numbers of science and engineering graduates produced. However there is slow improvement in high-tech manufacturing value added and a similar slow improvement in ICT expenditures with the EU/US gap cut by half since 1996. There have also been significant changes in levels of entrepreneurship across some member states.

Entrepreneurship in the European Union, some comparisons

The issue of levels of enterprise formation and attitudes towards enterprise was reviewed extensively in the EU Green Paper (Audretsch, 2003). Comparisons were made between people's attitudes towards entrepreneurs together with the number of actual entrepreneurs. Similarly, the dynamism in entrepreneurial activity and entrepreneurs' performance were compared. These give a varied picture with a preference for a more entrepreneurial society seen only in certain states and show marked differences with the position in the USA. In the USA, 67% of US citizens would prefer to be self-employed compared with an EU average of 47%. Across Europe the preference is to be employed rather than to be self-employed (Eurobarometer survey), although there is a relatively higher preference for self-employment in southern Europe, Ireland and the UK. Also the emphasis on high growth firms and job creation reviewed earlier is often associated with the entrepreneurial character of the owner, individual managers or operational teams (McCalman, 1999).

As might be anticipated, there is again a varied picture in terms of the incidence of entrepreneurs, including self-employed and business owners. In the US this is just over 10%, while across Europe, this varies widely from around 6% in Denmark and Luxembourg to 13% in Spain, 15% in Portugal and over 18% in Greece and Italy. Involvement in *new* entrepreneurial initiatives, shows similar trends; while in the USA 13% are currently taking steps to start a business, have set up a business or taken one over in the last three years, in the EU the average is only 4.5% of EU citizens are in the same position. Variation here shows 6% in the UK and Ireland to less than 2% in France. More than twice as many Europeans than Americans gave up their efforts to start a business. (Green paper, 2003)

US firms are, on average, smaller at start up than European firms, but successful firms grow more rapidly in terms of employment than their European counterparts. Hence US entrepreneurs seem to test the market by starting on a small scale and, expanding rapidly with success, whereas in Europe many business ideas do not reach the market as their viability is questioned even before they are tested in the market place. This aspect of risk aversion has been the subject of debate at national level, with the UK government setting up initiatives and measures to try to change attitudes to "risk" and "failure". However across Europe, there are higher levels of risk aversion in terms of business start up and development than would be the case in the USA.

Emerging models to empower innovation; SMEs and clusters

The first common factor across such models is their inclusion of dynamic and entrepreneurial small firms. Forming 908% of all firms across Europe such firms are expected to significantly contribute to both wealth creation and employment generation. There is a growing body of literature linking the dynamism of such SMEs to their location within industrial clusters, which are believed to engender innovation through frequent interactions and information flows (Antonelli 1999). Here clusters also need to be seen as changing organisms based on changing bonds and alliances. There is a view here that newer and smaller clusters will outperform older and larger clusters since although there are positive effects from cluster strength and new firm entry, these reduce over time as the cluster gets larger and older (Swann 1998). Similarly cultural homogeneity in the cluster may be problematic (Lazerson and Lorenzoni 1999), despite the views of earlier researchers (Piore and Sabel 1984). (World Bank, 2002).

However, both public and private investments and resource commitments are stimulated by the existence of regional and national clusters while a geographic concentration of suppliers and customers provides firms with shorter feedback loops for innovations. (World Bank 2002). Swann (1998) suggest that companies in clusters tend to grow faster than average; that clusters attract a disproportionate amount of new entry firms, that companies in clusters tend to be more innovative and that in some sectors, the strength of the science base in a cluster was correlated with higher than average new company formation and growth.

Locality may be vitally important, however, as earlier research has shown that the returns linked to the transmission of knowledge are geographically bounded (Jaffe, Trajtenberg and Henderson, 1993) and that this place-boundedness is also related to firm size. Firms with higher resources devoted to R&D -generally large firms- tend to rely on spillovers from distant institutions more often than firms with few or no resources in R&D – basically SMEs – which rely on local research institutions (Beise and Stahl, 1999). Proximity and location are thus important factors to be considered in the knowledge production function (Feldman, 1994; Audretsch and Feldman, 1996b); factors which ultimately lead to the spatial clustering of innovative activity, especially in industries which are R&D intensive (Audretsch and Feldman, 1996a) (Rodríguez-Pose and Refolo, 2002).

These advantages in turn stimulate the growth and persistence of localized clusters of firms. Porter's views have been faithfully put into practice across regional development agencies when he states that spatial proximity brings competitive advantage if the firm has to manage a complex set of interdependencies with clients, suppliers and other institutions (Porter 1990; Porter 1998). Clusters are seen in different industries and are "heterogeneous in size and pattern of relationships among firms" although they often they share some attributes, such as their large SME base. An example of this is the international competitiveness of Italian industries being mainly coincident with industries having SMEs clustered in industrial districts and regional networks (World Bank, 2002; Porter 1990). In addition, the literature also see industrial clusters as key drivers in terms of knowledge creation, since companies in a cluster share and create knowledge via face-to-face interactions and via the creation of common languages and institutions. If uncertainty and complexity are high, and iterative problem solving needed to ensure effective new product development then clustering enables the process (Solvell and Zander 1998).

The impact of industrial clusters on innovation centres around the issue of knowledge. This emphasis on knowledge has also been suggested by those proposing a replacement of geography related to locale and transportation costs by geography of organizational knowledge (Spender i1998). While explicit knowledge can be easily transferred through both old and new technologies, the transfer of tacit knowledge, strongly embedded in practice and often incommunicable, is typically accomplished by repeated interaction among individuals belonging to different organizations (Nonaka and Takeuchi 1995). It is the sharing of tacit knowledge and hence the empowering of innovation that is empowered by geographic proximity although now the use of more radical ICT solutions may supercede the need for proximity. The Digital Business Ecosystem Model being developed as part of the Framework VI programme uses the development of an open source software platform as a vehicle for empowering small firm specific software development, countering the disaffection expressed by many small firms about the software currently available to them (Martin, 2004).

If these new dynamic clusters are to succeed, what conditions are thought to be necessary for their progress? Such loose associations of innovative firms work closely with each other and with local universities and research institutes, business support and regional development agencies. Policy therefore needs to be designed and implemented which enables public sector bodies to support existing clusters and to encouraging new cluster formation in order to empower innovation. Hence not only the companies must be innovative in this process, this calls for public sector policies and actions to be similarly innovative.

Where regions have a more traditional industrial base, with an emphasis perhaps in difficult or declining markets – this Amy include agriculture as much as it does coal extraction given the pressures of the global market then different measures may apply. While those with high technology clusters may expect to create new knowledge applications, those in low technology clusters may need policies to encourage them to adopt new technologies and to be able to begin the process of knowledge development. The current focus on university enterprise in the UK with resulting patents, spin-offs and enterprise across all levels of graduates and staff, also depends upon the existence of a high knowledge base and high-tech clusters.

Data issues in EIS compilation

As has been indicated earlier, there are issues around the types of data collected and the comparability of such data. In some areas there is clear indication of this. As one example, the EIS states that comparing EU with US trends in lifelong learning is difficult, as the US does not collect such data. In addition, there are issues around the size of company, which forms the basis for comparison. In collating and analysing figures for the "small and medium sized enterprise" for instance, although the majority of SMEs across the EU comprise less than 7 employees, the EIS focuses on firms of twenty employees and above. There are clear reasons for doing so relating to availability of data etc, but the impact of focusing on such firms is readily apparent.

In addition, there is the issues of whether a consistent focus on high tech patents is useful overall; although the knowledge economy is characterized as having a high technology basis, the rates of innovation for mid and low technology aspects may also be important. Where innovation relates to industries not seen as high tech, will these also be included? New member states have traditional knowledge bases related to traditional industries, agriculture etc. If they are to measure their rate of change and development, it may be more useful to look at rates of change in this context, rather than adopt measures related to other types of economies or societies.

Other factors – questioning the internal focus

This paper will not consider the direct impacts of the enlargement process on innovation as this is considered elsewhere. However the over-focus on European trends etc can lead to a blinkered approach in considering how innovation may be affected by external factors. There is a range of such factors in the task environment for small firms involved in innovation. For national innovation the same range of external impacts must also apply. As an example this paper reviews some trends observed elsewhere globally which may impact directly on EUY innovative performance. This is not intended to be a thorough review of the task environment but just identified some aspects occurring worldwide which impact on European competitiveness.

a) Aspects related to the nature of the European Union itself, here an EU imposed restrictions – The use of English and of member languages
b) Aspects related to externalities elsewhere – The rise of Asian economies
c) Aspects related to regulatory differences – IAS 38
d) EU effects on others being reflected back – Changes in monetary systems reflecting the formation of the Euro

a) EU imposed restrictions – The use of English

The use of English as a language for business may seem to be a strange choice for entry into this debate. However, while China and India focus on use of English as the main business communication method the European Union has retained separate language policies. The enlargement of Europe does not reduce the confusion of translation and interpretation surrounding each EU communication, decision and meeting. American commentators have predicted this to be a limiting factor in business development, with resulting additional costs of translation of documents into a range of new languages as well as the standard German, French etc. compounded by the additional time taken to achieve these goals.

While large multi nationals adopted English as their business 'lingua franca' in the last decade, the EU retains the multi-lingual approach. Yet this is seen as an article of faith and an embodiment of the European ideal (encompassing diversity) rather than a potential barrier to business development. If this potentially damaging issue cannot be discussed in business terms, how can it be addressed?

b) Externalities – The rise of Asian economies

The rise of Asian economics has long been heralded and anticipated. Despite some setbacks in the 1990s for Asian tigers, there has been resurgence in Japan, and the seemingly inexorable rise in the influence and performance of the Chinese economy continues. If foreign direct investment flows into these economies and production transfers to these countries how will R and D be affected in the EU? Newsweek (2004) predicts an inflow of US investment in new member states rather than Asian economies, however the effects of either scenario need to be evaluated to predict effects on innovation and on economic development.

c) Regulatory differences – IAS 38

One major issue not so far explored is the effect of intangible assets on the balance sheet. Intangibles comprise those vital knowledge areas described above. On the balance sheet they may appear as negative quantities, representing cash spent rather than investment made. Despite debate the introduction of new accounting standards worldwide – International Accounting Standards (IAS) – will continue to treat the intangibles in this same way, as costs damaging to the balance sheet.

Signatories to this agreement include the EU and many developing countries. The USA, Canada and Australia are so far retaining their own system of amortizing intangibles, which allows costs to be deferred over time. With the need for companies to make profits in the short term, all too readily observable in European economies the pressure will be on tore duce costs. R and D, training etc will all appear on the "wrong" side of the balance sheet and may deter innovation. Will this be an unforeseen enabler for US innovation compared with that seen in the EU?

d) EU effects on others being reflected back – Changes in monetary systems reflecting the formation of the Euro

One new aspect in recent years is the introduction of the Euro. Whatever its direct effects on the European economy, it has had major indirect effects elsewhere with other groupings of countries wishing to emulate the idea of a single currently providing a larger internal market. One such has already been part financed by the Japanese government as research process – the ASEAN dollar. Feasibility studies are due to report back on the initial idea this year (Agarwal et al, 2004). The EU does not live alone in the world but affects and is affected by other factors.

Clearly these are quite arbitrary examples, selected to demonstrate the danger of relying of measures, which are too inwardly focused. There are issues too with the way in which data is used and collected for the existing measures – given comparability of data for instance. The impact this has can be clearly illustrated by the caveat on the SME tables, since although SMEs may be defined across Europe s 10-249 or even sometimes as 2-249, the survey only covered enterprises with 20-249 employees. Also the figures for education and for SE graduates may also be unreliable. In the words of the report

> "The indicator is a measure of the supply of new (S&E). Due to problems of comparability for educational qualifications across countries, this indicator uses broad educational categories. *This means that it covers everything from graduates of one-year diploma programmes to PhDs.* A broad coverage can also be an advantage, since graduates of one-year programmes are of value to incremental innovation in manufacturing production and in the service sector".

Conclusions

In a perfect world, it would be easy to identify winners and lose amongst nations, differentiating by some clearly discernible output called performance. In the real world however, the measures available offer some clue to the current state of European innovation. In this case the EU needs to work quickly and intensively to address key issues such as the rates for patenting, in order to achieve a better comparison with the US. However, in assessing. such performance care needs to be taken to include external risk factors, which may also impact on the analysis, and to agree more comparable data measures across member states. The emergence of new models to encourage innovation is welcome, and will provide the subject for much further research as Framework VI continues and Framework VII evolves.

References

Audretsch D. (2003), Green Paper; Entrepreneurship in Europe, Brussels, Document based on COM (2003) 27 final, 21 January

Acs, Z.J. / Audretsch D.B. (Editors), (1993), Small Firms and Entrepreneurship: An East West Perspective (Cambridge University Press, Cambridge).

Acs, Z.J. / Audretsch D.B. / Feldman, M.P. (1992), "Real effects of academic research: comment". American Economic Review 82, 363-367.

Acs, Z.J. / Audretsch D.B. / Feldman, M.P. (1994), R&D spillovers and recipient firm size. Review of Economics and Statistics 76, 336-340. Arundel,

Agarwal A. / Wong W.K. / Martin L.M. / Penn, J. (2004) "Asean Dollar: A Common Currency Establishment, for Stronger Economic Growth of Asean Region" Journal of Indian Finance, Delhi, May

Boari C. (2001), "Industrial Clusters, Focal Firms, and Economic Dynamism: A Perspective from Italy", World Bank

CII, CITT, IIFT, DSIR (2004), "Technology financing in SMES for international competitiveness" February

Dixon, T. / Thompson, B. and McAllister, P. (2002), "The Value of ICT for SMEs in the UK: A Critical Literature Review", Report commissioned by the UK Small Business Service, The College of Estate Management, September

Drucker P. (2001), "Innovation and entrepreneurhsip"

Forster, N. (2000), "The potential impact of third wave technologies on organisations", *Leadership and Organisational Development*, vol 21, issue 5, MCB

Grant, Robert. "Shifts in the World Economy: The Drivers of Knowledge Management" in Charles Despres and Daniele Chauvel (eds.) "Knowledge Horizons, The Present and the Promise of Knowledge Management" 2,000 p. 27

Iammarino, S. / Santangelo, G. (2000), Foreign direct investment and regional attractiveness in the EU integration process. European Urban and Regional Studies 7, 5-18.

Jaffe, A.B. (1989), Real effects of academic research, American Economic Review 79, 957-70.

Jaffe, A.B. / Trajtenberg, M. / Henderson, R., (1993), Geographic localization of knowledge spillovers as evidenced by patent citations, Quarterly Journal of Economics 108, 577- 598.

Martin L.M. (2004), "Computer programmers are from Mars and Small Firms are from Venus; the current status of international e-readiness in SMEs", EurIndia 2004 conference, New Delhi, March

Martin, L.M. (2003), "Valuing intangibles in the knowledge economy", keynote address Valuing intangibles in the Knowledge Economy Conference, Indian Institute of Foreign Trade and Indian Association of Chartered Accountants, March 7

Martin, L.M. / Matlay, H. (2001), "Blanket approaches to small firm internet adoption", Internet Research Policy and Practice, 11, 5

McCalman, J. (1999), Self managing teams in high technology manufacture: overcoming technological barriers, *Microelectronics International*, vol 16, issue 1, MCB: Bradford

Newsweek (2004), "Evaluating enlargement", April 30

Nonaka, I. / Takeuchi, H. (1995), *The Knowledge-Creating Company: How Japanese Companies Create the Dynamics of Innovation*, Oxford University Press, New York, NY

OECD (2002), "High growth SMEs and employment", Paris

Perez-Bustamante, G. (1999), "Knowledge management in agile innovative organisations", *Journal of Knowledge Management*, vol 3, issue 1

Poon, S. / Swatman, P.M.C. (1997) "Small business use of the Internet, findings from Australian case studies", *International Marketing review*, vol 11, issue 5

Raeside / Walker, J. (2001), "Knowledge: the key to organisational survival", *The TQM Magazine*, Volume 13 Number 3 2001 pp. 156-160. MCB University Press

Rodríguez-Pose, A. / Refolo, M.C. (2002), "The Link between Clusters of SMEs and Public and University Research in Italy" Department of Geography and Environment, London School of Economics, Houghton St, London WC2A 2AE

Santarelli, E. / Vivarelli, M. (1997), From which source do small firms derive their innovative inputs? Some evidence from Italian industry. Review of Industrial Organization 12, 243-258.

SIBIS (2003), "Measuring the Information Society in the EU, the EU Accession countries and the US", empirica GmBH

Stansbridge, P. / Shelton, R. / Sage, J. / Martin, L.M. (2003), "Knowledge Management Nade in Europe, vision 2010", report for the European Union

Swann, P. (1998) The Dynamics of Industrial Clustering, Blackwells, London

World Bank (2002), "Industrial clusters in Italy", http://www.worldbank.com/wbi/publications/wbi37186.pdf

Appendix: A European Innovation Scoreboard

The EIS indicators are divided into four categories, containing 17 main indicators and 3 additional Indicators for the Candidate countries, all selected to summarize the main drivers and outputs of Innovations:

- **Human resources for innovation, comprising 5 main indicators:**
  - 1.1 S&E graduates (‰ of 20-29 years age class)
  - 1.2 Population with tertiary education (% of 25-64 years age class)
  - 1.3 Participation in life-long learning (% of 25-64 years age class)
  - 1.4 Employment in medium-high and high-tech manufacturing (% of total workforce)
  - 1.5 Employment in high-tech services (% of total workforce)

- **The creation of new knowledge, comprising 4 main indicators of which two are divided into EPO and USPTO patents:**
  - 2.1 Public R&D expenditures (GERD – BERD) (% of GDP)
  - 2.2 Business expenditures on R&D (BERD) (% of GDP)
  - 2.3.1 EPO high-tech patent applications (per million population)
  - 2.3.2 USPTO high-tech patent applications (per million population)
  - 2.4.1 EPO patent applications (per million population)
  - 2.4.2 USPTO patents granted (per million population)

- **The transmission and application of knowledge, comprising 3 main indicators which are divided between manufacturing and services:**
  - 3.1.1 SMEs innovating in-house in manufacturing (% of manufacturing SMEs)
  - 3.1.2 SMEs innovating in-house in services (% of services SMEs)
  - 3.2.1 SMEs involved in innovation co-operation in manufacturing (% of manufacturing SMEs)
  - 3.2.2 SMEs involved in innovation co-operation in services (% of services SMEs)
  - 3.3.1 Innovation expenditures in manufacturing (% of turnover in manufacturing)
  - 3.3.2 Innovation expenditures in services (% of turnover in services)

- **Innovation finance, outputs and markets, comprising 7 main indicators of which three are divided between manufacturing and services:**
  - 4.1 Share of high-tech venture capital investment
  - 4.2 Share of early-stage venture capital in GDP
  - 4.3.1.1 Sales of 'new to market' products in manufacturing (% of turnover in manufacturing)
  - 4.3.1.2 Sales of 'new to market' products in services (% of turnover in services)
  - 4.3.2.1 Sales of 'new to the firm but not new to the market' products in manufacturing (% of turnover in manufacturing)
  - 4.3.2.2 Sales of 'new to the firm but not new to the market' products in services (% of turnover in services)
  - 4.4 Internet access/use
  - 4.5 ICT expenditures (% of GDP)

4.6 Share of manufacturing value-added in high-tech sectors
4.7.1 Volatility rates of SMEs in manufacturing (% of manufacturing SMEs)
4.7.2 Volatility rates of SMEs in services (% of services SMEs)

R&D Priorities in Innovation Policy and Financing in Former Socialist Countries
W. Leal Filho and P.S. Gramatikov (Eds.)
IOS Press, 2005

Chapter 3
An Analysis of some Aspects of Innovation Processes in Armenia

B. Gabrielyan, T. Arzumanyan, A. Khosrovyan
National Academy of Science of Armenia
24-D Bagramian, N 907, 375019; Yerevan
Armenia

Summary

This chapter introduces the state of innovation in Armenia and current developments at the country level.

Introduction

For all the former USSR countries one of the crucial problems is "how to wean scientific research away from dependence on government support and make science and scientists learn to survive and thrive in the private business sector. This problem must be solved quickly to prevent the erosion and degradation of the national science base" (Turner, 1996).

In the post-Soviet period, in many countries such as Armenia the vast majority of the scientists and technologists has not been given a part to play. It is important to prepare a field for their activity in order to preserve the intellectual resources of the country. Many international organizations have provided collaborative and other grants to support the existing intellectual potential in Armenia, as well as making direct investments to found new companies. Owing to Armenia's investment policy, which is one of the most open in the post-Soviet era countries, the government has made significant efforts to attract foreign investment to Armenia. As a result, there are currently over 2,000 partially or fully owned foreign companies in Armenia, including several multinational companies such as Alcatel, Lycos, VirageLogic, Marriot International and Coca-Cola.

There are many programmes aimed at developing innovative technologies thus enlarging a network of technological centres all over the world and promoting technology exchange between nations. However, many research activities still cannot be completed in Armenia because of the lack of long-term financial support.

It is known that the S&T and Innovation system comprises three main components:

- Generation of S&T knowledge
- Transmission of S&T knowledge
- Application of S&T knowledge

Another important condition is the availability of adequate financial organizations ready to finance this process.

Let us consider the situation in Armenia with reference to the above-mentioned aspects.

According to the data of Armpatent, during 1993-2001 the total number of filed inventions and useful models amounted to 1684 in Armenia, including 1217 by national and 467 by foreign entities. On the basis of these applications, 1077 patents for inventions and 43 for useful models were issued. By comparison, during 1980-1990s 5293 patents were issued to national entities for inventions and useful models in Armenia. Based on 2001 population census data it can be calculated that the current index (2.8) for the average number of patents for inventions and useful models per 100,000 of population is around 5.7 times lower than the corresponding index for 1980-1990, which was 16.1 (Ghazaryan, 2002).

As for technology transfer and application institutions, during recent years several technology parks were established which are starting to be one of the key factors for future economic success.

For example, Enterprise Incubator Foundation is a non-profit business development incubator founded in 2002 by the Government of Armenia and funded by a loan from the World Bank within the "Enterprise Incubator" initiative. It aims to assist local IT and high-tech companies and stimulate economic growth in these areas: driving Armenian technology products to international markets, promoting the managerial and professional capabilities of local technology companies, assisting with the formation of new technology start-ups and their further development and attracting foreign direct investment to Armenia.

Open Source Armenia is the first project in Armenia in the IT area aimed at developing a community of IT professionals and showcasing their potential and accomplishments to the world. Initiated at the beginning of 2003, it is intended to provide new marketing channels and business models. Several large-scale projects are already hosted in the domain opensourcearmenia.com, where any interested company or individual can use free software or further customize it for special needs.

Another success story, Viasphere Technopark is a wholly owned subsidiary of Viasphere International, a Silicon Valley based venture capital firm and incubator that invests in start-up technology companies and provides management services to them. Viasphere invests in companies in the US, then helps these companies establish subsidiaries in Armenia by providing the office space and infrastructure to enable the subsidiaries to work with their parent companies in the US.

In 2001, Viasphere International purchased and renovated one of Armenia's former semiconductor plants to build Viasphere Technopark. The Technopark is a 10,000 square metre facility that offers office space and state-of-the-art networking, communications and computer infrastructure to technology companies. It already hosts several technology companies – among them more than 10 software and hardware companies which were established with the help of Armenian professionals from America's Silicon Valley. According to future projections this number will grow to 20, showing that the Technopark can play a central role in promoting technological innovation and collaboration in Armenia.

In 2003, the American company Monterey Design opened a research and development centre at Viasphere Technopark in Yerevan, Armenia, under the name of Monterey Arset. The centre employs over 30 senior scientists with an electrical engineering or computer science background. The centre plans to "aggressively expand its activity in Armenia in future", according to Jacques Benkoski, president and chief executive of Monterey Design.

Another technopark has been created, on the base of two institutes in Armenia, which hosts the science-based technologies of several companies – synthesis of cosmetic products, production of bioadditions including powder "Narine" and production of pure copper cable and wood slats for furniture. It is planned to set up production of small-size trucks and construction techniques on the basis of components supplied by Japan.

Unfortunately, current legislation sometimes even hinders development and activity of the technoparks, such as inadequate customs rules. These rules do not differentiate the type and weight of the products of scientific value as compared with high weight trade ones.

According to some practitioners, success in the development and implementation of innovative technologies depends very much on the effective state system of management of intellectual property, which includes:

a) the legislative base, such as laws on patents, trademarks etc.;

b) independent state institutions dealing with patent and antimonopoly issues, and corresponding structures in the Ministries of Justice and Internal Affairs and the State Customs Committee;

c) a team of patent attorneys and evaluators of intellectual property;

d) an agency to deal with intellectual property rights on behalf of the state and monitor the control over the use of such rights;

e) economic incentives resulting from using intellectual property including tax discounts, and an appropriate credit policy to promote the use of innovative technologies.

Currently, only the legislative base is more or less developed in Armenia, including the "Law on Patents", "Law on Copyright", "Law on Trademarks" and several others. Of the independent state institutions mentioned above, ArmPatent was operating until March 2002. Currently, ArmPatent together with the Agency on Copyrights has been amalgamed in the newly created Agency on Intellectual Property, which is a structure within the Ministry of Trade and Economic Development of the RA. Drafts of various laws relating to intellectual property issues have been prepared and submitted, however; until now they were not addressed with proper attention by interested structures (Kantarjian).

With the purpose of promoting innovative practices in the country, a Government Commission for developing suggestions on governmental regulation and control over the export of scientific and technological production and intellectual property was established in 1993. However, the commission's activity only resulted in a few unrealized suggestions. Currently, no alternatives to the above-mentioned commission are present in Armenia.

Economic incentives are rather poor, and no reliefs or privileges related to the provision of credits for the commercialization of the products of intellectual property are being considered in the legislation of the RA.

In 2003, the Ministry of Trade and Economic Development considered preconditions for the commercialization of developments. An "Technology Transfer" association, comprising several institutions, was created for this purpose. The association is cooperating with analogous foreign organizations. It also provides consulting in the field of innovation to Armenian enterprises. The association is intended to help such enterprises to acquire international innovative products, and to serve as an intermediary in the international innovation business.

According to official data, in 2000-2001 the volume of realization of scientific developments was increasing by 50% compared to each of the preceding years. In 2002 it was only 15%, which is an indication that the old management practice has run out of steam. New approaches to foster the further development of innovation in Armenia are required (Khandanian, 2004).

Another problem is the inadequate level of cooperation between local business entities. Local businesses are not sufficiently aware of the developments taking place in Armenia. For instance, only recently one of the firms has started to buy the cheaper local production instead of importing it from abroad. One of the reasons for such a passive use of local production by the local businesses is in the fact that there are no guarantees that local scientific institutions will continue to exist in the long term. Anyway, it is obvious that our science-intensive institutions have yet to learn how to sell their production in the country.

Below are presented some of new and old innovative technologies developed in various Armenian institutions, in different stages of implementation.

In the *Institute of Biochemistry* of NAS Armenia, long-term investigations (over more than 40 years) on the human and animal hypothalamus have been conducted, and biologically active compounds have been discovered. As a result of the investigations, neuropeptide bio-regulators were discovered which have a large spectrum of pharmacological activity. Among them is the proline-rich polypeptide. Experimental testing on animals has revealed that it has pharmacological effect:

- regulating the reaction of cell and humoral immunity, stimulating the activity and functions of immune system cells and many other properties;
- antibacterial effect:
- increases viability, adhesia and accumulation of phagocytes, and strengthens their bacteriocidic properties, inhibits bacteria reproduction and development, etc.;
- antiviral effect: inhibits encephalomyocarditic virus replication and increases gamma-interferon level in blood cells with mitogen-stimulated nuclei.

As an immunomodulating remedy it can be used in cases with septic processes, post-operational complications, burns, mixed infections of the genital-urinary system, irradiation damages, against inhibited blood formation after irradiation and chemotherapy on patients with tumours, etc. It has shown favourable properties as compared other immunomodulating remedies (levamisol, bestatin, taftsin, etc.). To be tested fully the remedy should pass through the steps prescribed by World Health Organization (WHO). Unfortunately, financial support is required and further investigation has been suspended.

In the *Institute of Radiophysics*, a new type of powerful solar power station has been developed. Its advantages over existing ones are:

- one fixed mirror instead of numerous moving mirrors;
- ground area occupied for an order (0.7 ha instead of 7 ha) for a station providing 1-1.5 megawatts of power;
- a new type of heat exchanger with high efficiency;
- low cost (more than for order) and higher efficiency (more than 15 times).

Many of those improvements have been patented already.

The prototype, which produces a power of 100 KWt, has been designed; for its implementation a financial investment of 1.5 million dollars is required.

The *Institute of Mining and Metallurgy* has invented a method of extracting selenium from acid solutions. Copper and molybdenum concentrates contain a significant portions of noble metals – selenium, rhenium, silver, gold, etc. In the course of processing, the concentrates rhenium and selenium enter into acid solutions The addition of natural bentonite powder to these solutions causes selenium absorption to take place. This is a cost-effective way of removing selenium in an intermediary step of the metallurgic treatment of the molybdenum concentrates, compared to the traditional method of selenium sublimation from slag, using an expensive technological process. This development has been partially tested in one of the plants of Vladikavkaz, and copyright has already been obtained. Further cooperation on this development would be welcomed.

Armenia's high-tech and information and communication technology (ICT) industry is represented by a wide variety of private companies, associations and NGOs, R&D institutions, higher learning institutions and training centres:

According to data from the high-tech/ICT portal www.siliconarmenia.com:

- 54 software development companies
- 24 hardware development companies
- 12 ISP/telecom companies
- 2 consulting companies (2000)
- 7 NGOs /associations

Most of them were started after 1990, and the current favourable foreign investment policy in Armenia is encouraging an increase in their numbers. The portal itself was developed with a grant from USAID Armenia, SETA Corporation and the World Bank. Its mission is to support the creation of an ideal business environment to attract additional foreign investment in the Armenian high-tech/ICT industry, promote that industry in the high-tech global market place and advance the creation of a knowledge-based economy in Armenia.

The most successful software and hardware companies operating in Armenia are listed below.

Synergy International Systems Inc., is a US-based software company, a leader in the development of web database applications for e-government and business intelligence solutions using advanced Intelligent Data ManagerTM (IDM) technology. Its mission is to give large public and private sector organizations the rapid ability to create and manage complex and dynamic websites. In 2000 the company's management established an office in Yerevan.

LEDA Systems, Inc., founded in 1995, is a leader in the field of mixed-signal and analog intellectual property design and development. LEDA provides a single comprehensive source for mixed-signal IP, including specialized I/O components, DAC/ADC and PLLs. LEDA has licensed its technology to more than 50 customers, including Agilent Technologies, Intel and LSI Logic. LEDA's primary product lines are: Analog SIP, Special I/Os and Product IP for the communications market.

Virage Logic Corporation is a leading provider of embedded memory products. It licenses its intellectual property, including embedded memories, memory test processors, compilers to semiconductor companies and integrated device manufacturers. According to Adam Kablanian (Chairman, President and CEO of Virage Logic), the company provides the blueprint which is designed to be integrated into a chip. In other words, it provides the intellectual property and works in close association with chip manufacturers. Among its customers are ATI Technologies, AMCC, Intel, TranSwitch and Vitesse Semiconductor, Toshiba. In 1996, understanding the importance of high-performance memories for semiconductor companies, the first subsidiary was established in Armenia.

Conclusions

Armenia possesses considerable innovative potential and certain positive shifts in the field of R&D and innovation are obvious. To ensure the successful continuation of the process, however, it is necessary to specify what we want and how to achieve it. For this purpose it is very important to study and apply the best results of economic experience, generally accepted in the world, taking into account of course the specific features of our country. This calls for the in-depth analysis of the innovation activities of advanced countries at the present time and for the past 30-40 years, and of the appropriate innovation development models.

It is also essential to develop and enforce the relevant mechanisms of state regulations in science, technology and innovation, including a complex system of state support for small innovative businesses. In Armenia some legal acts referring to this field have already been passed. But still there is no law "On innovation and innovative activity", and the existing ones do not provide real mechanisms directed to promoting R&D and innovation, particularly with reference to the development of an innovative SME sector. This can be expressed through "tax vacation" and reliefs for SMEs involved in R&D, and determining a certain percentage of state orders to be placed with SMEs.

Government should ensure that individuals and organizations continue to have strong incentives and opportunities to capitalize on research. Based on international experience, one of the possible effective mechanisms could be national "sectoral" funding programs which are intended to promote high-quality research and development in a given industrial sector, through establishing special funds in which a percentage of corporate taxes are targeted to funding innovative R&D activities. Such funds will require close interaction of the indigenous academic community, the private sector and government with a view to creating it, setting its priorities, and managing it.

References

Ghazaryan, L. (2002) "Statistical Analysis of Patents in the Republic of Armenia", Apricot Plus Ltd., Yerevan.

Inventing a better future: A strategy for building worldwide capacities in science and technology. InterAcademy Council, 2002

Kantarjian, S. (2004) "Innovation Process is Impeded due to Lack of Legislative Base", Delovoy Exspress Weekly, N7, Yerevan.

Khandanian, A. (2004) "Technoparks are to be Organized on the Base of "Andron" and "Chimtekh" Institutes", Delovoy Exspress Weekly, N3, Yerevan.

Science and Innovation as Strategic Tools for Industrial and Economic Growth, Ed.C.Corsi, NATO ASI Series, Science and Technology Policy, Vol. 4, 1996.

Shirak International Association, Innovation Technologies, Yerevan, 2003

Open Source Armenia web site www.opensourcearmenia.com

Hi-tech/ICT portal web site www.siliconarmenia.am

R&D Priorities in Innovation Policy and Financing in Former Socialist Countries
W. Leal Filho and P.S. Gramatikov (Eds.)
IOS Press, 2005

Chapter 4
CORDIS: Drawing from Experience of European Support for Innovation

PETER WOLFMEYER
ZENIT GmbH
Dohne 54
D-45468 Mülheim an der Ruhr
Germany

Summary

Innovation is a top priority for Europe's policy makers and essential for sustainable economic development and employment, and thus also ultimately for the standard of living and quality of life of each and every European citizen. It is crucial that the research and business communities understand European R&D and innovation policy and strategically orientate their scientific and commercial activities towards its objectives. Since its inception, CORDIS has supported this process by providing a comprehensive information and assistance service.

Introduction

European industry is characterised by an insufficient level of innovation and a widespread failure to transform novel ideas into marketable products. In addition, compartmentalisation still prevails to a major degree, that is, RTD and innovation activities are taking place in isolation and without coordination in individual EU Member States (in separate "compartments"), which often also leads to duplication of both activities and expenditure. How can this be combated ? Large-scale endeavours have been and are being undertaken to establish transparency in EU RTD activities, one of which is CORDIS, the **Co**mmunity **R**esearch and **D**evelopment **I**nformation **S**ervice.

This paper positions CORDIS and its role in the Sixth Framework Programme as well as amongst the most important partner programme lines. It also presents its key features and the results generated by an analysis carried out by the Innovation Relay Centre at ZENIT of data concerning participation in EU-funded projects by organisations in North Rhine-Westphalia.

ZENIT: The one-stop shop for information and advice

ZENIT, the Centre for Innovation and Technology in North Rhine-Westphalia, has its finger on the pulse of innovation and technology development in the region and beyond. Tracing, tracking and turning into reality new innovations and technology trends is ZENIT's core activity.

ZENIT Zentrum für Innovation und Technik in NRW

WHO WE ARE

ZENIT Zentrum für Innovation und Technik in Nordrhein-Westfalen GmbH
founded in 1984 as a Public Private Partnership
to support the structural change in North Rhine-Westphalia

Founded in 1984, ZENIT GmbH is a non-profit company which provides consulting services to corporate and institutional clients in future-orientated fields of technology, innovation management as well as in policy approaches and evaluation. The team of experts covers disciplines ranging from microelectronics and mechanical engineering to empirical social research and economics. ZENIT's target groups comprise first and foremost industrial enterprises, institutions and public authorities in North Rhine-Westphalia. Numerous consulting projects have been carried out in the areas of

- market research
- technology screening
- adaptation of employees to new technologies
- change management, training
- regional development strategies
- S/T policy design and implementation
- programme evaluation
- technical assistance.

ZENIT's services are also sought in the field of financial engineering where financial concepts are developed and implemented by combining various components, i.e. public funding, corporate loans or venture capital.

Activities range from publications and seminars targeted at the general public to tailor-made consulting services for individual firms and public organisations. Since ZENIT primarily focuses on SMEs, most of its services are geared to their needs. Clients benefit from ZENIT's close involvement with a multitude of institutions, initiatives and organisations and from its status as a public private partnership, since its recommendations on SME issues are often sought by the respective ministries.

ZENIT hosts a Euro Info Centre which was founded in 1987 as one of the pilot EICs. As SME specialists, its aim is to inform, advise and assist SMEs in North-Rhine Westphalia in all Europe-related areas. The EIC has access to all the relevant European Commission databases, including CORDIS.

Thanks to ZENIT's participation in the EIC, OPET and last but not least the Innovation Relay Centre (IRC) networks, it is familiar with the modus operandi and working environment of the European Commission. Not only on behalf of clients but also for itself, ZENIT has successfully placed several proposals for funding under European schemes – be they part of the RTD framework programmes or otherwise. Several studies have been carried out on behalf of the European Commission and a large number of projects implemented.

North Rhine-Westphalia: powerful on an EU and global scale

ZENIT's activities cover the geographical area of the Federal State of North Rhine-Westphalia (NRW), Germany's most densely populated region. Bordering on Belgium and the Netherlands, NRW has some 18 million inhabitants and 30 cities with more than 100,000 inhabitants (almost 22% of Germany's overall population on only 9.5% of its total land area). NRW has 6 international airports and a dense network of motorways and railways. NRW is Germany's strongest export state: 18% of goods exported from Germany are "Made in NRW".

In economic terms, NRW is a very important region not only in Germany but also in a European and global context. It is one of Europe's ten most important regions in terms of scientific innovation and home to the highest concentration of technology centres/incubators (69) and research centres (23, including 10 Max Planck institutes and 8 Fraunhofer institutes) which guarantee dynamic research and development activities. 4,300 people are employed at Jülich Research Centre alone. There are seven "Networks of Competence" in the Aachen region, where 10% of all German researchers work. These are bwA – Biomaterials Aachen, car e.V. – Competence Centre Automotive Region Aachen, Laser Technology Aachen, Medical Technology Aachen, Plastics Innovation Aachen, REGINA e.V. – Information Technology Aachen and SimPRO – Process Simulation Aachen.

5 of Germany's 10 largest universities are in NRW; about 500,000 students are enrolled at 53 universities and technical colleges. Know-how and new technological developments have helped secure this federal state's international competitive standing and position as a top-class business location.

NRW's trade fair status reflects its national and international importance. The state's four exhibition centres – Düsseldorf, Cologne, Essen and Dortmund – are meeting points for world business. Every year, about seven million guests visit the special trade exhibitions, as well as the trade shows open to the public. 40% of all the international fairs in Germany take place in NRW.

ZENIT Zentrum für Innovation und Technik in NRW

NRW

The marketplace in the heart of Europe

40% of the European Union's inhabitants live within a radius of 500 kilometres from Düsseldorf, NRW's federal state capital. They represent 45% of the total purchasing power in the European Union. The gross domestic product of NRW (2002: 464 billion Euro) represents 22% of the German GDP. 21% of Germany's overall working population works in NRW. In addition to the expanding service sector, which includes telecommunications and media, NRW boasts an impressive list of various industrial sectors involved in manufacturing and processing:

Key industrial sectors in NRW (figures for 2002)	Turnover (billion Euro)	Employees (in 1000s)
Chemical/pharmaceutical industry	43	128
Mechanical engineering	34	217
Metal production and processing	28	118
Food and beverage industry	27	102
Automotive industry	26	90
Paper, publishing, printing	17	92
Production of rubber and plastic goods	11	71
Textile and clothing industry	10	44
Measuring & control, medical technology, optics	3	25
Radio, TV and news technology	10	22
Wood processing	3	21

At global level, the Rhine-Ruhr region is seventh in the league of the most economically powerful areas after Tokyo, New York, Osaka, Los Angeles, London and Paris. In Europe it occupies third place, being surpassed only by London and Paris. North Rhine-Westphalia ranks 14th in the global list for GDP, ahead of the Netherlands (15), Australia (16) and Switzerland (17).

The Sixth Framework Programme (FP6)

By contrast with previous framework programmes which were characterised by a strong thematic orientation and, whilst they contributed to developing a culture of scientific and technological cooperation, did not make a really lasting impact on coherence at European level, the Sixth Framework Programme features a much stronger focus on mechanisms aimed at supporting the innovation process. The objectives of the Sixth Framework Programme are therefore

- to concentrate Europe's efforts on fewer priorities
- to move towards the progressive integration of activities of all relevant participants working at different levels
- to promote research activities designed to have a lasting, "structuring" impact
- to support activities which will strengthen Europe's general scientific and technological basis and
- to use the scientific potential of candidate countries to prepare and assist their accession to the EU to the benefit of European science at large.

These objectives stem from the recognition that research and development in specific scientific areas is insufficient to establish a true European culture of innovation, and to answer the problems Europe faces as a result of its lingering behind the USA and Asia. Of crucial importance is a coordinated and transparent environment which encourages research and innovation, avoids duplication of activities (and thus of costs) and which ensures the generation and transition of excellent results into marketable products and services which in turn contribute to a better quality of life for Europe's citizens (e.g. more and sustainable employment).

In comparison to past Framework Programmes, the Sixth Framework Programme is structured in only two main areas which answer two key priorities. The first area, Specific Programme 1 "Integrating and Strengthening the European Research Area (ERA)", is concerned with the "Thematic Priorities" of the Framework Programme and aims at more focussed and integrated research at EU level in seven fields of science and technology (Life Sciences, genomics and biotechnology for health; information society technologies; nanotechnologies, multifunctional materials and new production processes; aeronautics and space; food quality and safety; sustainable development, global change and ecosystems (including energy and transport research); citizens and governance in a knowledge-based society). 11,285 million euros are dedicated to this area. There are, of course, additional budgets for cross-cutting issues such as research in support of EU policies, multi-sector research activities involving SMEs, support for coordination between national and regional research programmes. The second pillar of the Sixth Framework Programme is Specific Programme 2 "Structuring the European Research Area" where there are four so-called "Activities": research and innovation; human resources and mobility; research infrastructures; science and society. An overall budget of 2,605 million euros is allocated to these priorities. The total budget for the Sixth Framework programme is 17.5 billion euros (including Euratom).

A key feature of the Sixth Framework Programme is the introduction of two new instruments: "Networks of Excellence" and "Integrated Projects". These instruments have been devised as a new method of implementation of the Framework Programme. Past Framework Programmes were implemented on the basis of cooperative research projects which demonstrated two weaknesses: the end of a research project generally meant the end of the consortium as well, and many projects failed to achieve the necessary critical mass to make a real impact either in scientific, industrial or economic terms. The philosophy behind the Networks of Excellence and the Integrated Projects is to move away from multiple project funding and towards the funding of coherent programmes of research activities. Networks of Excellence aim to integrate progressively the activities of network partners and thus create "virtual" centres of excellence. Integrated Projects are large-scale projects intended to establish the critical mass in objective-driven research with clear scientific and technological goals.

CORDIS and its partner programme lines IRC network, ETI and IPR Help Desk, which are presented in this paper, all fall under the Research and Innovation theme of Specific Programme 2 "Structuring the European Research Area".

The IRC Network

The first IRCs were established in 1995 with the support of the European Commission as members of an integrated, pan-European platform to stimulate transnational technology transfer and promote innovation services. Today, 68 IRCs cover a wider geographical area than any other technology transfer network in the world, spanning all the EU Member States as well as Bulgaria, Cyprus, the Czech Republic, Estonia, Hungary, Latvia, Lithuania, Malta, Poland, Romania, Slovakia and Slovenia, Iceland, Israel, Norway and Switzerland.

ZENIT Zentrum für Innovation und Technik in NRW

IRC

The IRC Network
http://irc.cordis.lu

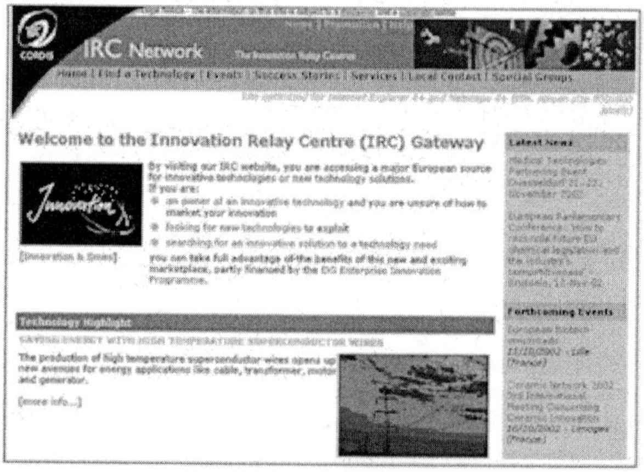

Most IRCs are operated by a consortium of regional organisations with established reputations in local research and industrial communities. These include innovation agencies, Chambers of Commerce, regional development agencies and university technology centres. Altogether, almost 250 partner organisations are involved, ensuring a local presence in most regions of Europe. IRC staff – a total of nearly 1000 – are experienced specialists with backgrounds in business, industry and research.

However, there are also single IRCs (i.e. not operating within a consortium), and a developing trend is that of transnational IRCs. Since April 2002, IRC NRW is also a transnational IRC (see below).

The mission of the Innovation Relay Centres is to facilitate innovation and transnational technological cooperation in Europe with specialised business services. These services are primarily targeted at technology-oriented small and medium-sized enterprises (SMEs), but are also available to larger firms, research institutes, universities and technology centres. Traditional as well as high-tech sectors are covered. In today's highly competitive global markets, SMEs frequently need to acquire technology from abroad, to sell or license technology outside their own country or to enter into joint development arrangements with foreign firms which have complementary know-how. However, many SMEs lack the information, resources or specialised skills to find suitable transnational partners and negotiate satisfactory agreements efficiently and reliably. In all these cases, the local IRC offers a single point of access to the entire range of expert support required – either delivering services in-house or signposting the client to other regional or European agencies.

IRC services include:

- assistance with technology acquisition and technology marketing
- support in identifying suitable partners for technology cooperation
- partner matching and technology brokerage events
- technology audit and technology watch
- help in negotiating technology partnership agreements
- access to the results of EU, Eureka and other research programmes
- access to EU innovation financing and IPR support services.

Each IRC's approach, procedures and skills have been developed in response to the technological structure and business culture of the region it serves and each has established close ties with its local community of technology-based firms. At the same time, all IRCs draw strength from their membership of a European network. The strong technical and personal links within the network equip each IRC to mediate technology transfers quickly and efficiently. As members of a European network, the Innovation Relay Centres connect their regions to innovative capacity in other countries. They enable local firms to source the technologies they need from the best available suppliers and provide them with a channel for their own know-how to new markets in Europe.

The process of innovation is complex and takes time. It requires access to research results, adequate means of protecting intellectual property, availability of development finance and a supply of appropriate business skills. And these elements must be coordinated to create an environment which encourages entrepreneurs to take risks. At regional level, IRCs are helping to create such an environment by offering entrepreneurs a single point of access to the entire range of expert support they need – both that supplied by their own staff and systems and through links to other regional and European resources. However, the IRCs' core business remains transnational technology brokerage. It is here that they have built their reputation and it is their services in this field which add the greatest value to the efforts of Europe's entrepreneurs and researchers.

The IRCs' transnational technology brokerage service is primarily targeted at technology-oriented small and medium-sized enterprises (SMEs). In today's intensely competitive and rapidly changing markets, most technology-based firms must pursue European and international sales from the outset and need foreign licensing, distribution or co-development partners at an early stage. At the same time, many must look outside their own regions for the technologies they require to improve their productivity or add value to their product. It is to streamline the international development of such SMEs that the IRC service is especially designed, although it is also available to larger companies, research institutes, universities and technology centres.

Each IRC maintains a portfolio of present and potential clients with whom it stays in touch through newsletters and local events in addition to specific technology transfer projects. Wherever possible, an IRC adviser with relevant technological competence is assigned to each client. Understanding and trust, based on personal contact, is often the prerequisite for a successful IRC project.

Critical for the IRCs' core task of transnational technology brokerage is a database of technology offers and requests which is maintained centrally by the IRC-IRE Central Unit in Luxembourg and accessed by the IRCs via a web-based intranet. A request for technology partnership submitted on behalf of a client company is distributed electronically across the network within hours – and may quickly produce suitable responses from enterprises in other corners of Europe. Every case is unique, and the IRCs' core technology brokerage service very often depends on extensive prior contact between the IRC and the client.

The Innovation Relay Centre NRW/Malta at ZENIT

The main mission of the Innovation Relay Centre is the promotion of innovation and transnational technology transfer. The figures presented above show that NRW offers considerable potential for the work of the IRC NRW in terms of R&D capability, innovation capacity and the sheer number of enterprises, initiatives, organisations and other players which make up the innovation and technology landscape in NRW. On the other hand, the IRC NRW has its work cut out to reach all these protagonists and to integrate them in the transnational technology transfer process – its fundamental mission. That is why the IRC NRW has developed two major strategies over the course of its evolution:

1. To present a clearly defined IRC profile to its clientele with special IRC services and expertise, yet in parallel integrating itself in the regional network of technology transfer and innovation.
2. To offer not only wide-scale actions targeted at larger audiences but also individual services tailored to the requirements of specific enterprises or groups of enterprises, i.e. clusters and thematic groups.

Being a highly innovative and technology trend-setting region, NRW tends to produce rather than require new technologies (i.e. is a donor rather than a recipient of new technologies). This is also reflected in the projects recorded in the CORDIS database, which we will examine in greater depth later.

A product which the IRC produces in cooperation with the Euro Info Centre at ZENIT is the "EU Information Service". The aim of this monthly publication is to filter out that information of relevance to ZENIT's clientele, i.e. primarily SMEs, from the vast volume of information and data available about EU topics. The newsletter covers calls for proposals, partner search profiles, calendar of events, descriptions of programmes from the Sixth Framework Programme and other initiatives. There is also a weekly "Infoticker" which is disseminated to subscribers via e-mail, giving an update of the week's information highlights. One of the main sources generating material for these services is CORDIS.

ETI

Another key priority of the Sixth Framework Programme which is also closely related to CORDIS is ETI. ETI is the acronym for Economic and Technological Intelligence. This is a vital component of competitive research and innovation strategies in a knowledge-based economy. The scheme commenced under the Fifth Framework Programme, where over 50 ETI projects were funded. ETI actions are a support measure for the participation of SMEs in European research activities, the aim being to improve their access to scientific and technological information. Under the Sixth Framework Programme, there will be a number of calls for ETI projects. Each project must address a specific research theme or industrial sector. The scale of these projects is flexible, with a budgetary range of 0.2 to 2 million Euro. The time scale envisaged for a project's implementation lies between 1 and a maximum of 3 years. Possible methods to support the goal of enhanced SME participation could, for example, be by facilitating the creation of clusters of SMEs with similar innovation needs, promoting transregional clusters between SMEs or stimulating networks of industrial incubators. ETI project contractors are intermediaries (e.g. SME National Contact Points, industrial federations, networks and associations of research performers, professional associations, chambers of commerce) and organisations experienced in economic and technological intelligence. Like the IRC project, ETI projects are classed as Specific Support Actions, may however also be Coordination Actions. An example of an ETI project is CETISME (**C**ooperation to promote **E**conomic and **T**echnological **I**ntelligence in **SME**s), in which a Spanish-led consortium with French, Italian and British partners analysed ETI activities in their regions with the aim, via a subsequent benchmarking process, of identifying best practice in SMEs. The project ran under the Fifth Framework Programme and was awarded nearly half a million Euro EU co-funding.

IPR Help Desk

The IPR Help Desk is another Specific Support Action within the Research and Innovation area of the Specific Programme "Structuring the European Research Area" which goes hand-in-hand with the IRCs, ETI and CORDIS. A study of the European Patent Office showed that two-thirds of enterprises generating inventions do not use the patent system. The purpose of the IPR Help Desk is therefore to create a better awareness and greater use of the system by facilitating access to various sources of information on Intellectual Property Rights (IPR) and to help project contractors – both potential and current – with all issues pertaining to Community diffusion and protection rules and intellectual property in transnational research projects. The organisation awarded the contract following the IPR Help Desk call under FP6 will be responsible for the following tasks: assistance, dissemination of information and training measures related to IPR issues relevant to Community-funded transnational research projects (e.g. licensing, technology transfer, consortium agreements), as well as the provision of specific tools such as a website, helpline, newsletter. The helpdesk performs an information rather than an advisory function and thus does not replace the assistance available from patent offices, agents or lawyers. It was developed in response to similar initiatives in the US and Japan.

CORDIS – An introduction

CORDIS, as the name implies, lies at the *heart* of all EU RTD and innovation activities and is the main information and communication platform. It is the authoritative information source on European innovation. CORDIS is part of a service package aimed at assisting the creation of a European-wide innovation system and contributes to counterbalancing national compartmentalisation by aiding the flow of information, the spreading of knowledge and the diffusion of innovation. The CORDIS service is free of charge and is characterised by easy navigation.

The objectives of CORDIS are threefold: to be a user-friendly, multilingual information dissemination service which facilitates participation in Community research and innovation activities, as well as to encourage the improved exploitation of research results with an emphasis on sectors crucial to Europe's competitiveness and to ensure a diffusion of knowledge, thus fostering the innovation performance of enterprises and the societal acceptance of new technology. CORDIS' target group is anyone interested in the European Union's RTD and innovation policy and actions, in particular the Framework Programmes, and it addresses parties seeking appropriate programmes, calls, projects or partners, offering not only information but also concrete assistance and step-by-step guidance on how to participate in and benefit from EU programmes: an Ariadne-type aid through the labyrinth of EU programmes and other initiatives. To give some idea of the scale: CORDIS comprises over 250.000 database records, 60,000 web pages organised into 179 web services and 50,000 downloadable documents. On average, up to eight million CORDIS pages are viewed each month; here the FP6 service is the one most frequently visited (up to one million pages viewed per month). Per month some 300,000 documents are downloaded. CORDIS is used in over 150 countries: EU Member States represent 82% of CORDIS users (Association and Candidate Countries = 9%; SMEs = 13%). The information available on CORDIS is provided by the European Commission services, external organisations, Member States, Accession and Associated Countries as well as regional authorities. Users are research organisations, SMEs and large companies, national administrations, policy makers, intermediaries, support organisations and journalists.

CORDIS is structured in five main areas: research programmes and other initiatives, innovation policy, horizontal/database-related services, national and regional gateways and information support services. Clearly the volume of data available within CORDIS cannot be presented within the scope of this paper, therefore the author has selected a few examples of CORDIS services of particular pertinence to his audience.

CORDIS services: some examples

European Innovation Scoreboard (EIS)

The European Innovation Scoreboard was born as a result of an explicit request of the European Council of Ministers meeting in Lisbon in March 2000. It annually assesses innovation performance in the EU Member States. As the same indicators are used in each country, for example, the number of science graduates or the number of patents filed, the scoreboard is a benchmarking tool which can be used to compare countries, especially since the EU average is also given, as are the equivalent figures for Japan and the USA. Since it also includes historical data, it gives a picture over time of trends in innovation performance. The scoreboard provides a large volume of very detailed information, for example, country performance, innovation in regions, innovation in services, R&D versus diffusion, innovation versus GDP, technology sectors. Data are given for all EU Member States, the Accession and Candidate Countries, the Associate States and other countries (USA and Japan), thus permitting numerous ways of analysing the figures. Policy makers thus have an instrument that enables them to identify the strengths and weaknesses of their country and to take appropriate action, e.g. by looking at the strategies and methodologies of countries which are succeeding. The most important aspect of the scoreboard is however not the individual figures but the comparison of them and the disparities they reveal.

Technology Marketplace

CORDIS is not just about access to funding, it is also a gateway and a platform for the evaluation, promotion and exploitation of results. The Technology Marketplace is a tool to support the transfer of results into the marketplace. The user can browse through a large number of offers which are categorised into four areas: business, science, society and FP6 research themes. The latter also provides an opportunity to see what research has been performed in the past so that anyone planning a new research proposal can find experienced partners and build on past work, rather than overlapping.

ZENIT Zentrum für Innovation und Technik in NRW

CORDIS

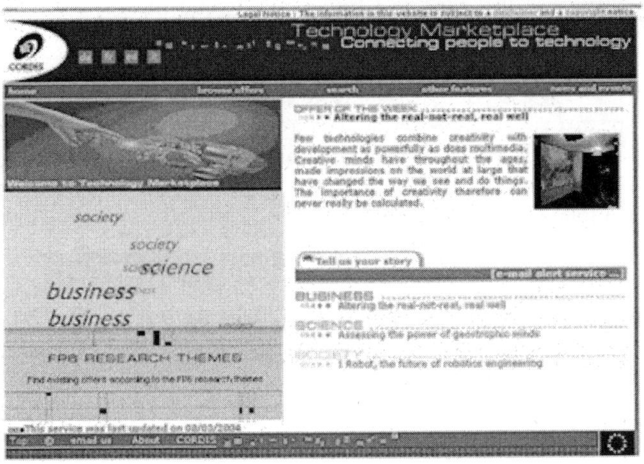

www.cordis.lu/marketplace

The QuickPick service is an automatic alerting service which updates the user on new offers on the basis of key words. Using the search mechanism, final stage results and prototypes can be identified which are ready to be transformed into marketable products. The interactive service is updated weekly and appears in five languages.

CORDIS Partners

The CORDIS Partner Search Service covers all EU countries, EFTA and EEA, pre-accession and other countries and allows the user to publish and update his own profile online, as well as to search for suitable partners on the basis of the profiles they have submitted. In the Success Stories section, examples are given of partnerships mediated via the service.

CORDIS Express

CORDIS Express is the weekly newsletter to which users can subscribe. It is a useful means of familiarising oneself with CORDIS since it provides a synthesis of the new information published in the various CORDIS services and databases. Using QuickPick e-mail alert, readers can receive automatic notifications on new calls and follow all CORDIS updates on their desired topic as well as on events, projects, partners and research results.

Regional Gateways and NRW gateway

The origins of the CORDIS Regional Gateways go back to early 1998 when the Italian region of Piemonte set up a website entitled “Focus on Technology and Innovation in the Piemonte Region”. In the summer of the same year the service was opened up to other regions. The demand was so great that in December 1998 a pilot action was launched: "Focus on Research and Innovation in the Regions Service", whereby a platform was established for individual regions to promote their research and innovation-related infrastructures on CORDIS. In January 2000 the latest step in this development followed with the start of the CORDIS Regional Research and Innovation Service, the key objective of which is to "provide a dynamic information work space for regional research and innovation activities, focussing on local and region-based actors" (e.g. SMEs, universities). The information workspace itself has three core objectives: to enable regions to promote their local research and innovation environment, activities and infrastructures, to provide local actors in the participating regions with a tangible interface to EU research and innovation activities and to provide information and links to other EU-supported resources, initiatives and services which help local actors to benefit from EU research and innovation activities.

To date, 28 regions from 10 EU Member States (Germany, France, Italy, Spain, Denmark, The Netherlands, Finland, Eire, Austria and the UK) have entered their presentations on the website, including North Rhine-Westphalia.

Our presentation gives an insight into our regional technology policy and the highly developed innovation and technology landscape in North Rhine-Westphalia. The main features of NRW's technology policy are

- Close cooperation between all regional players (enterprises, higher education institutions, research institutions, associations and unions)
- Creation of clusters and formation of networks, i.e. interplay of all those involved in industry, technology, education and politics
- Faster transformation of globally available knowledge and information from HEIs into the products and production techniques of local enterprises
- Increased use of the support programmes offered by the European Union and the federal government
- Establishment of technology and industry initiatives.

In addition, the NRW Regional Gateway has links to CORDIS and to other regional players such as the North Rhine-Westphalian Ministry of Science and Research, ZENIT, research institutes as well as other initiatives and projects such as Euregio. Two further areas are descriptions of key regional projects and news about latest developments in the fields of research and innovation in the region and on forthcoming events. ZENIT contributes to the regular updating of the Gateway by supplying important technology news from the region to CORDIS for integration into the website.

CORDIS Correspondents

In 2003, Directorate General Enterprise, and more specifically the "CORDIS Marketing" unit, reinforced the CORDIS Correspondents action. Each Member State was asked to nominate specialists for this role. The author of this paper was appointed CORDIS Correspondent for Germany. The role of a CORDIS Correspondent is threefold: In an advisory capacity, the CORDIS Correspondent shares his knowledge about the needs of CORDIS end-users in his country. He is the "eyes and ears" of CORDIS. The CORDIS Correspondent also promotes CORDIS nationally by helping users to understand better and benefit from the various CORDIS services. In his role as multiplier, the Correspondent ensures there is an interaction between CORDIS and national activities, e.g. he communicates CORDIS' benefits to national players at national events. The main tool employed in the day-to-day activities of the correspondents is the CORDIS Correspondents Bulletin Board which is a communication platform, the so-called "Online Forum" between the Correspondents and CORDIS management within the Commission as well as between the Correspondents themselves. Some of the Correspondents are also National Information Providers (NIPs). In addition to the bulletin board, CORDIS Correspondents Days take place once a year. The last one took place on 8 September 2003 in Luxembourg and addressed the subject of how to improve CORDIS.

CORDIS Analysis

Over a longer period, the IRC North Rhine-Westphalia has analysed the data available from CORDIS in an endeavour to identify those local SMEs in the region which have participated in the 5 completed Framework Programmes to date, i.e. covering 18 years from 1984 to 2002. This data has been supplemented by our own searches. The analysis includes all the Specific, Thematic and Horizontal Programmes as well as programmes outside the Framework Programmes such as THERMIE, LIFE, the education and training programmes and the e-commerce programmes.

ZENIT Zentrum für Innovation und Technik in NRW

CORDIS

CORDIS Analysis: Structure

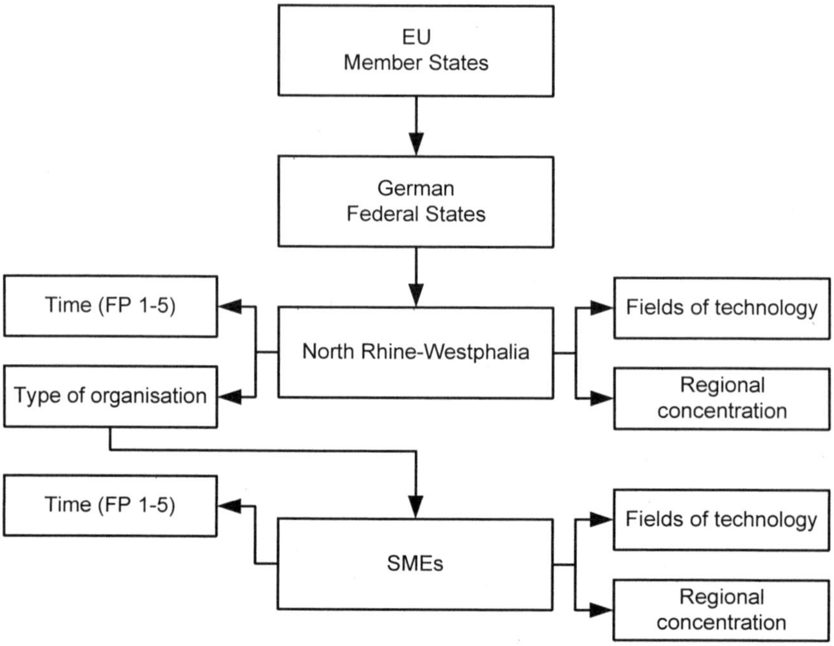

Our analysis is broken down into a number of levels, examining first of all the overall participation of all EU Member States, then comparing North Rhine-Westphalia's participation with that of the other German federal states and also studying what types of organisation, i.e. SMEs or universities/research institutes, have been involved in the different Framework Programmes. In total, there were 15,978 projects from the EU15 in which 19,004 German partners were involved. There were 5,006 projects with at least one partner from NRW. As some projects included more than one NRW partner, the number of partners is obviously higher than the number of projects, i.e. 5,802. This does not however mean that 5,802 different organisations participated. Often the same organisations have been involved in various projects. The number of individual organisations is 1,358 of which 587 were SMEs (43%).

Another aspect of the analysis is the question which areas of technology have dominated. For North Rhine-Westphalian SMEs, the areas of Competition and Growth, Information Society and Energy and Environment were the most important under FP 5.

Particularly illuminating is the distribution of participation within North Rhine-Westphalia, with the area around Aachen clearly dominating. This is reflected not only in the high level of participation by research institutes and higher education institutions but also by SMEs, indicating that the interplay between the two types of organisation and the general innovative climate in the area (we recall the several Networks of Excellence clustered around Aachen) have stimulated a high degree of interest and participation in the Framework Programmes.

The main conclusions which ZENIT was able to draw from the analysis were that

- SMEs now recognise the advantages of participating in EU projects. This is reflected in the constantly growing number of organisations involved.
- North Rhine-Westphalia, together with Baden-Württemberg and Bavaria, plays a major role in EU projects. The other German federal states are far less significant.

Conclusions

From experience, it can be stated that CORDIS is an excellent source of information and assistance. It provides a wealth of data to all players in Europe concerned with research and innovation. Not only are the contents permanently updated, CORDIS itself is also continuously evolving and being constantly developed and refined so that under FP6 CORDIS means "new, state-of-the-art, value-added information and communication services".

Chapter 5
An Overview of the Main Issues in Innovation Financing in Ukraine

OLHA KRASOVSKA
Centre for Scientific and Technological Potential and Science History Studies
(STEPS Centre), National Academy of Sciences of the Ukraine
60 Taras Shevchenko Blvd., 01032
Kiev 32
Ukraine

Summary

This chapter provides an analysis of the main issues that concern and influence innovation financing in Ukraine, with a critical appraisal of the matters which need to be considered in order to improve current trends.

Introduction

Traditionally in the market economy financial institutions resources are one of the main source of innovation financing. Unfortunately in Ukraine such a system of innovation activity support is not well developed. In particular only 6% of innovation activity of the industrial enterprises is financed by different credits (Fig. 1). This circumstance makes the development of a mechanism for attracting financial institutions to innovation financing extremely important.

Main tendencies in the financial institutions sector

Nominally in Ukraine the sector of indirect financing, which implies participation of different financial institutions, is developing rather fast. This may be proved by the fact that there were 630 investment companies and funds and 1407 auditor firms by the end of 1996 in Ukraine, while at the beginning of 2000 there were already 1330 investment companies and funds and 2250 auditor firms functioning. Besides in Ukraine 157 commercial banks and more than 300 insurance companies are registered. (Varnaliy et al., 2002 p. 23; Kozhevina et al., 1997, p. 84)

Capabilities of different financial institutions to take part in innovation activity

The potential opportunity for financial institutions to take part in the innovation activity is determined by two major factors:

- Their ability to use what is known as "long money", which then may be a source to give credit to innovations; and
- State regulation of the investment activity of financial institutions.

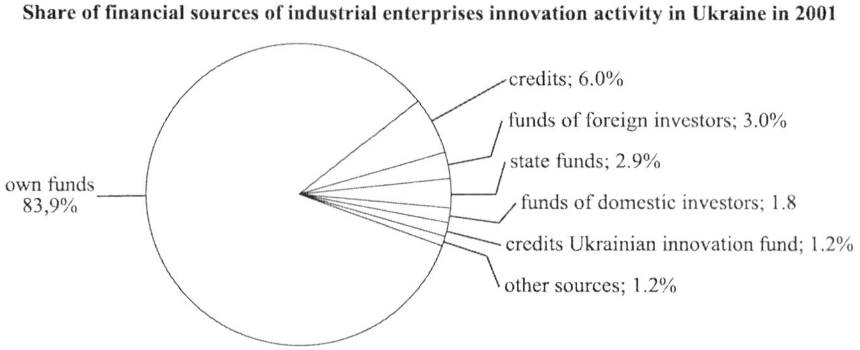

Fig. 1: Distribution of financial sources for innovation activities at industrial enterprises in Ukraine in 2001 (Science, 2002)

Commercial banks as a rule make decisions concerning their investment portfolio independently, while the investment activity of insurance companies and pension funds in majority of countries is limited by state regulation. Investments of these financial institutions are subject to rigorous state legislative and administrative regulation that consists of:

- General investment principles (principle of safety, profitableness, liquidities, principle of investment portfolio diversification etc.);
- Qualitative investment indicators (regulation of directions where investment are made);
- Quantitative restrictions concerning the volume of investments.

In particular in Germany the investment activity of insurance companies is controlled by the Investment Ordinance for insurance companies, passed on January 1, 2002. This ordinance cancels §54a of the Insurance Supervision Act and provides a certain liberalization of existing investment conditions for the insurance companies. For example, in this ordinance more attention is given to the foreign investments of insurance companies, and it also contains more favourable conditions for investments in the real estate and investment certificates. Simplification of conditions for investment in securities on the part of private enterprises is another positive trend with reference to the investment activity of insurance companies regulated by this ordinance (German, 2002).

Further the scale and scope of attracted financial resources and crediting will be described briefly. Judging by these indicators, Ukrainian financial institutions have good capabilities (potential) to be active participants in innovation crediting. At the same time, it should be noted that, for example, insurance companies considerably concede to the commercial banks in the scale and scope of attracted financial resources.

At the beginning of 2003 the cumulative assets of the insurance companies were more than 16 times less than the assets of the bank system, their authorised capitals were almost 4 times less than those of banks, and their own equity capitals was 3 times less. Thus despite the nominal increase of the main indicators of the insurance market, it is characterized by low scales of authorised capitals, equity capital, assets and reserves.

In Ukraine the investment function of the insurance remains underused because of:

- Insignificant scales of the financial resources accumulated by Ukrainian insurance companies;
- Impossibility of forecasting the moment of insurance cases coming;
- Complicated schemes of insurance for customers;
- Low level of management in insurance companies;
- Great number of financial reporting documents.

All these factors lead to an irrational use of reserves by the majority of insurance companies, which in its turn prevents an objective assessment of the real scale and scope of the Ukrainian insurance market (Baranovskiy, 2003).

In Ukraine different types of life insurance, which are the main source of long-term investment resources and, accordingly, a financial source of innovation crediting in the future, are not very popular today. Assets of Ukrainian financial institutions are invested mainly in bank deposits, securities and bonds. Clearly such types of investments could not be the sources of innovation crediting. That is why, actually, the majority of Ukrainian financial institutions rather prefer speculative forms of crediting, which provide them with "fast money", to the financing of innovation projects, which are long-term and highly risky. Nor does Ukrainian legislation satisfy the requirements of innovation investments.

Ukrainian commercial banks and SMEs as the agents of innovation financing

There are also a number of problems relating to the innovation statistics of financial institutions. For example, credits classifications used by commercial banks do not register credits for innovation separately. The analysis of the Ukrainian credit market shows positive trends. First of all it concerns the long-term credits in relation to the total amount of the credits. This share has increased from 12.4% in 1991 to 28.3% in 2002. (Bulletin, 2003) This can be an indirect indication of the shift of crediting priorities from manufacture to innovation projects. But the actual level of innovation crediting is lower than the level of long-term crediting as part of the long-term credits go to non-innovation projects.

Some commercial banks do use the term "innovation" in their reports, but with the implication that the bank is using innovation in its own activities and does not refer to credits for innovation projects. Despite the significant scale of credit operations by commercial banks in Ukraine, the share of innovation constitutes only a small part of their credit portfolio.

Nowadays small and medium sized enterprises (SME) are active agents of innovation activity. In Ukraine official statistics do not gather information on the level and characteristics of innovations in SMEs. The results of the international research project 'Innovations, Small and Medium Sized Enterprises and Economy Development in Ukraine and Belarus' (INTAS-99-0943) was used by the author to bridge the gap. According to these findings SMEs have serious financial problems but only few of them use external financial sources. Today informal credits of family and friends, credits of customers, suppliers, other enterprises, leasing etc., are the main sources of SMEs' innovation financing (Smallbone et al., 2002).

Venture capital as the form of innovation financing in Ukraine

The venture capital sector in Ukraine is only at the development stage. There are 3 venture funds in Ukraine: Western NIS Enterprise Fund, Euro Ventures and the "Ukraine" Fund. The limited number of funds is limiting creation of a competitive environment. All these funds are foreign and have invested in the Ukrainian economy about USD 100 million, or two USD dollars per inhabitant, while in Western Europe this indicator is about 50-60 USD.

For comparison purposes, according to the data of Russian Venture Investment Association in Russia venture funds collected 2.5-3 billion euros, and by the end of 1998 they had invested 375 million euros in 125 companies. In Ukraine the scale of venture investments equals 1% of annual investments of European venture funds, and just 1-2% of private enterprises get venture financing in Ukraine (Denisjuk & Krasovska, 2002).

Innovation legislation in Ukraine

In order to create an adequate legislation system in Ukraine some laws need to be developed and adopted. First of all, this concerns the law "On venture financing", as the law "On institutions of joint investment", which is supposed to regulate the venture sphere, does not cover all the peculiarities of venture financing. The law "On venture financing" should define the main terms in this sphere (venture fund, venture company, venture financing etc.), interrelations between the basic actors of venture activity, issues of creation of the positive fiscal and legal environment for venture activity, issues of risk reduction and stimulation of venture funds activity, tax privileges for venture investments, the mechanism of providing venture finance to innovation and high-tech SMEs venture financing, especially at the seed and early growth stages etc. Adoption of the law is only part of the issue of creating favourable conditions for venture sector development. Besides this, legal practice also calls for adjustment to the new situation by foreseeing certain stimuli and sanctions.

Respectively the general tax policy should have a stimulating character in relation to the venture capital industry; have to include certain privileges for venture funds and the innovation companies. The procedure for the licensing and registration of venture funds should be significantly simplified. In general the legislation system in the innovation sphere should be based on the Ukrainian Law "On innovation activity", the concepts of science and technology innovation development of Ukraine (these laws in Ukraine are developed but do not work now).

Conclusions

From this paper it can be concluded that the innovation-financing system in Ukraine is not relevant to the market economy needs, and its further improvement demands, first of all, an adequate legislative system and the creation of effective stimuli for financial institutions, which may and should be actively involved with innovation activity in the country.

References

Baranovskiy, A. (2003). Big problems of small market, Zerkalo nedeli, No 30 (455), August 9, p. 10.

Bulletin of National bank of Ukraine (2003). No 9, pp. 65-150.

Denysjuk, V. and O. Krasovska (2002). Venture financing in EU countries and possibilities of its use in Ukraine, Problems of Science, No 11, pp. 9-21.

German Insurance Industry (2002), Yearbook, Gesamtverband der Deutschen Versicherungswirtschaft e.V. (GDV), Press and Public Relations Department, Berlin. http://www.gdv.de/download/GDV StatYearbook 2002 navig.pdf

Kozhevina, N. & D. Lyapin (Eds) (1997). Small business in Ukraine. Entrepreneurs are able to revive the country, Association entrepreneurship development assistance "Ednannya", Kiev.

Science and innovation activity in Ukraine (2002). Annual statistics book, Kiev.

Smallbone, D. et al. (2002). Innovations, Small and Medium Sized Enterprises and Economy Development in Ukraine and Belarus. Final report of the international research project (INTAS-99-0943).

Varnaliy, Z., O. Sergiyko & L. Khmelevska (2002). Small entrepreneurship. In: Monitoring, infrastructure, personnel, Institute of private legislation and entrepreneurship of Legal Academy of Sciences of Ukraine, Kiev.

Chapter 6
Innovations in Poland

PIOTR PRZEWROCKI, JOANNA KULCZYCKA
MEERI PAS
Ul. Wybickiego 7
31-261 Kraków, Poland

MARIAN JASKUŁA
Chemistry Department
Jagiellonian University,
ul. R. Ingardena 3
30-060 Kraków, Poland

ALINA SIONKOWSKA
Faculty of Chemistry
Nicolaus Copernicus University
Ul. Gagarina 7
87-100 Toruń, Poland

Summary

This paper presents the Polish innovation policy, with a new direction within the subject outlined by the EU accession process. It also shows cases studies of cooperation between industry and academia in innovation, development and implementation.

Introduction

In the Information Society the main production factor is knowledge, which partially replaces capital. The main basic sources for knowledge are Research and Development works, and knowledge itself is the source of innovation processes towards products, technology, business organisation and human resources improvements (Goliński 2003). Innovations usually are defined as projects and construction works of services and products (both material and non-material); purchase of technology; instrumentation with industrial and engineering works; pre-industrial development works; launching production and marketing of new products/services (OECD – KBN 1999).

Development of Polish science is in the control of the government, in which strong divisions between sectors is noted. This is connected with the budget distribution between different ministries. Responsible for financing science in Poland is the Ministry of Science – Chief of State Committee for Scientific Research. For financing economic developments, responsibility is held by the Ministry of Economics, Work and Social Policy. For financing education, the Ministry of National Education and Sport is responsible.

Education in Poland

Based on the latest Statistical Yearbook of Central Statistical Office of Poland from 2003, more than 10 million inhabitants have completed secondary education (among 31 million population older than 15 years). There is also a significant rise in the number of people who possess higher education degrees. In 1988 this figure was 1.8 million, while in 2002 3.2 million had obtained a higher degree. 7.5 million people have a professional degree. That type of education is not presumed, however, to be adequate preparation for the needs of contemporary economics. On the other hand, just one in four people in Poland has finished elementary education (8.8 million people). 1.5 million inhabitants over 15 years old do not possess any formal education and/or qualifications.

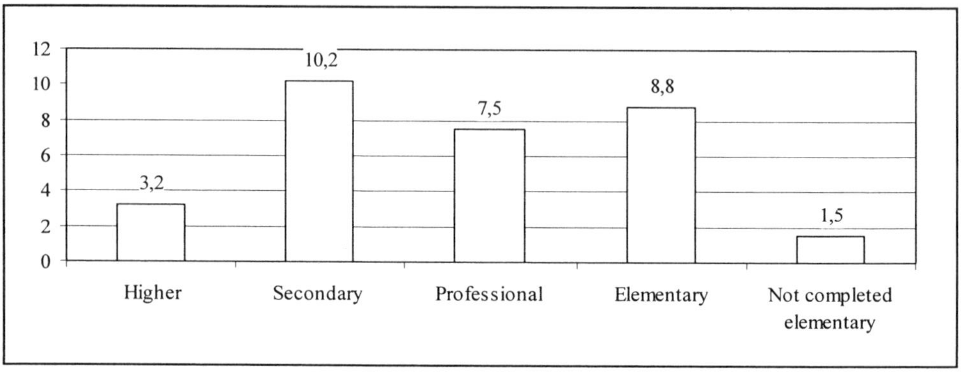

Fig.1: Education in Poland (Data for 2002 in millions).
Source: Central Statistical Office, Concise Statistical Yearbook 2003.

Human resources and infrastructure employed in R&D

In 2001 an increase in the number of companies with their own R&D laboratories and innovations departments was noted. However, the scale and impact of this new undertaking is rather insignificant owing to low R&D expenditures and the number of personnel employed.

The number of privately owned higher education facilities and universities which conduct R&D work increased to 19.

Type of unit	Amount	Employed staff	Direct R&D staff
Total	920	78 026	56 918
Polish Academy of Sciences Units	81+23*	6 934+281*	4 646+195*
R&D units	232	21 708	12 210
Higher Education Units	121	43 180	36 597
Development Units	463	5 923	3 270

Fig. 2: R&D employment (2001)
Source: Central Statistical Office. Concise Statistical Yearbook 2003, *including data for 23 independent research units within PAS.

The total amount of people employed in R&D activities in Poland was 78,026 in 2001. Staff directly employed in R&D works came to 56,918 people. The highest employment was noted in Higher Education Units (both private and state owned) – 36,597. Research and Development units employed 12,210 people, while Development units had 3,270.

In recent years a decrease in the total number of staff employed in R&D activities has been noted. This was caused by a significant decrease in the number of additional staff employed, while the number of researchers increased.

The ratio of researchers employed to 1000 inhabitants in 2001 was 3.3 (similar to Greece and Portugal). The highest ratio was noted in Finland (13.1), while the EU-15 average was 5.4.

Current innovation issues in Poland

One of the main innovation indicators which is used to assess innovation incentive is the relation of total Research and Development expenses to the Gross Domestic Product of the country. Figure 3 shows that in recent years a gradual decrease in R&D spending has been noted. Since 1999 when the relation of R&D expenditures reached the level of 0.75%, gradually it has been reduced in the following years. In 2000 the level of the ratio was 0.67%, then it diminished to 0.65% in 2001, and eventually in 2002 reached 0.59%. This tendency is caused not only by the lack of increase (or decrease, quite often) in R&D expenses, but also by the increase in GDP (5% a year on average).

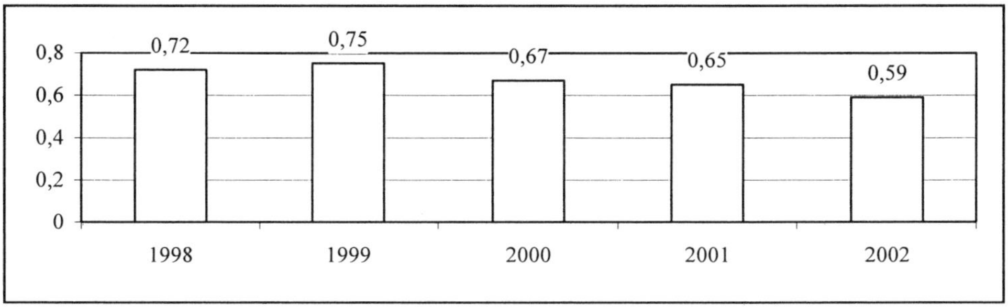

Fig. 3: R&D and GDP relation in Poland (%)
Source: Government Strategy 2004.

The above figures and tendency are rather alarming. According to the Lisbon Strategy from 2000, European economy should be more innovative, competitive, dynamic and modern and should become a knowledge-based economy. One of the main objectives in order to achieve such a goal is to reach an average level of 3% of expenditures in relation to GDP by the year 2010. It is highly improbable that Poland will be ready to provide such a level of involvement in the EU's innovation strategy.

	Thousand of millions USD	Percent of GDP
Sweden	9.9	4.3
Finland	4.7	3.4
Japan	103.8	3.1
South Korea	22.3	3.0
USA	282.3	2.8
Germany	53.9	2.5
Denmark	3.8	2.4
average for OECD	65.4	2.3

	Thousand of millions USD	Percent of GDP
France	35.1	2.2
Great Britain	29.4	1.9
Holland	8.4	1.9
Czech Republic	2.0	1.3
Hungary	1.3	1.0
Poland	2.6	0.7
Greece	1.1	0.7
Turkey	2.7	0.6

Fig. 4: R&D expenditures in selected countries (2001)
Source: Main Science and Technology Indicators, OECD 2003.

Poland, with R&D spendings in relation to GDP of 0.7%, is not only far behind the leaders in this ranking (Sweden with 4.3% GDP, Finland with 3.4% GDP, Japan with 3.1% GDP) but also it is much worse than other former Communist countries. For example, the Czech Republic spent about 1.3% of GDP on R&D, while Hungary spent 1.0%. Similar levels of R&D expenditures to those in Poland are noted in Greece and Turkey.

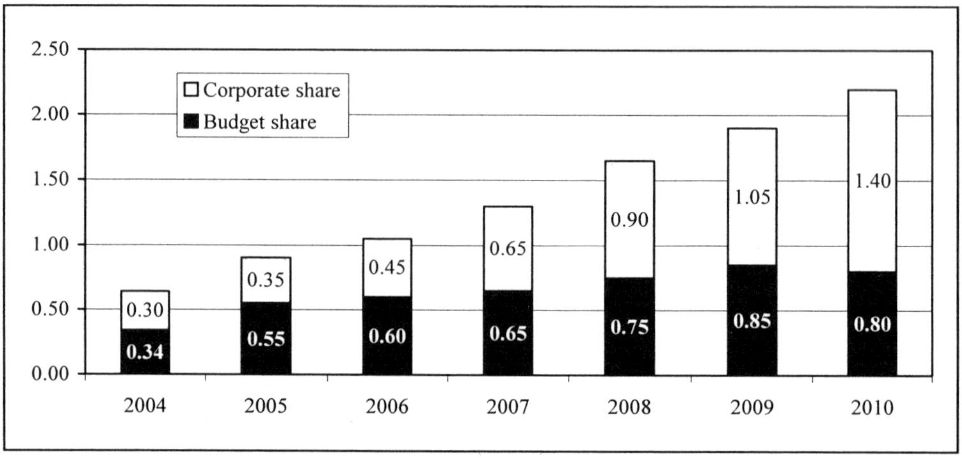

Fig. 5: Increase in R&D expenditures according to optimistic governmental scenario
Source: Government strategy 2004.

According to the very optimistic governmental forecast, total expenditure on R&D will grow steadily up to the year 2010. This increase is mainly driven by the rise in corporate spending on R&D. That amount grows from 0.3% of GDP in 2004, through to 0.65% in 2007, finally reaching 1.4% in 2010. At the same time governmental expenditures on R&D are also growing but more stable. In 2004 governmental expenditure is estimated at 0.35% of GDP. It grows to 2005 (0.55%), 2006 (0.6%), 2007 (0.65%), 2008 (0.75), 2009 when it reached maximum with 0.85%. In 2010 that amount diminishes to 0.8%.

Figure 6 shows the corporate incentive for financing R&D in selected countries. The overall incentive measured in that way is not satisfactory in Poland, and reached 35.8% of total expenditure on R&D. Only two countries are financing their R&D from the governmental budget to a greater extent – Turkey and Greece. At the other end of the scale, countries like Sweden (with 77.6% of R&D financed by corporations), South Korea (76.2%), the USA with (74.4%) show a completely different attitude towards financing innovations. It is presumed that money spent by companies directly is more effective and is spent on issues with more practical applications.

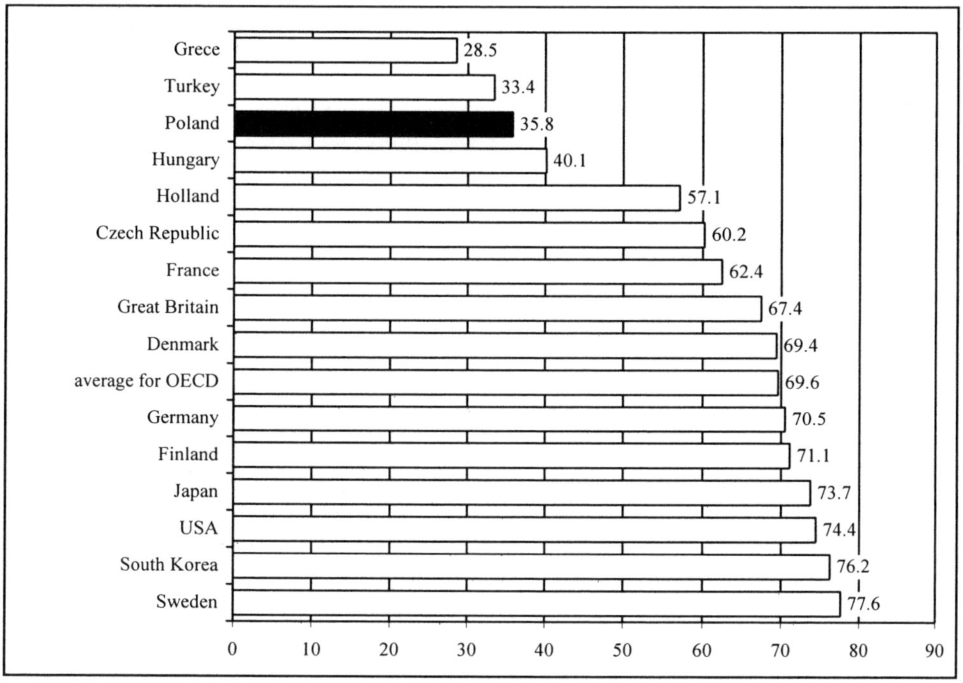

Fig. 6: Corporate participation in national R&D expenditure (Data for 2001 in %)
Source: Main Science and Technology Indicators, OECD 2003.

Intellectual property

From 1989-2002, the number of application patents filed dropped significantly (from 5294 to 2313). In 2001 the smallest number of application patents was filed since the beginning of transition period (2202). Figure 7 presents the trend line of this development. The trend is rather typical for countries during the transition period, and illustrates adjustments to market economy needs.

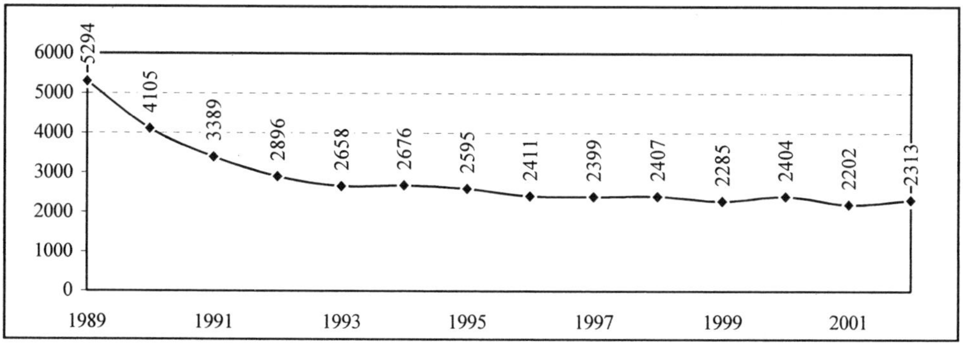

Fig. 7: Patents applications filed
Source: Central Statistical Office. Concise Statistical Yearbook 2003

In 2000, the highest numbers of patent applications were filed by the residents of Poland in the following countries: Germany (150), Austria (146), Spain, Switzerland (145 each),

Sweden (144), France (103), the Netherlands, Italy (102 each) and the United States (100). 102 patent applications were filed at the European Patent Office (EPO).

In 2000, the numbers of patents granted to Polish applicants were highest in the following countries: the United States (13), France (11), Germany (9), the United Kingdom (8), Italy (6) and Austria (5). The European Patent Office granted Polish applicants 7 patents (CSO 2003).

Foreign trade and high-tech production

One significant factor for economic competitiveness is the proportion of highly technological products in a country's exports. Highly technological products are those in which the scientific and innovation level is significant (drugs, computers, biotechnology etc.). During the years 1993-1999 the proportion of highly technological products in the country's exports was at the level of 2-3%. This was one of the smallest among candidate countries. A significantly higher share was noted in the Czech Republic, Estonia and Hungary.

The total share of high and middle to high technological industry products in exports is estimated to be 30%, while in some countries (Mexico, USA, Switzerland, Ireland) it is higher than 70% (Japan more than 80%) (MSRIT 2003).

Sources of innovation financing in Poland

According to the publication "Innovation policy issues in six candidate countries: The Challenge", Poland and Hungary were the only candidate countries which offer indirect support for innovation in companies by the use of tax benefits. There are however many direct ways in which R&D financing is supported from both EU and domestic sources.

The main sources of direct financing innovation in Poland are:

- Pre-accession funds: especially PHARE programs provide effective sources for financing innovative undertakings in companies. These funds will be probably be available until 2006. Subsidies can reach 50% of the total innovation cost. This source can finance both innovation consulting and direct investments. Financing is available for SMEs for the most part.
- Structural funds which provide companies with a rather attractive form of financing innovation activities. Directly supported R&D work performed for companies by research institutes and universities, subsidies buying patents and new technologies. The level of financing may vary from 30% to 100% of total qualified costs. Financing available both for SME and larger companies.
- Funds available for SME's by participation in the 6th European Framework Research Programme.
- Offset – opportunity of acquiring, among others, knowledge and technology from American companies and research institutes, as a compensation programme for weapon purchasing. Total direct offset liabilities are estimated at 740 millions of USD. This support is available mostly for large companies.
- Environmental innovations can be co-funded by the National Fund for Environmental Protection and Water Management. The level of financing may vary, depending on whether the undertaking is profitable and if money is required by a municipality or company.
- The Ministry of Scientific Research and Information Technology can provide companies with money for performing joint R&D undertakings with research institutes/universities. This work should have strong practical implications. The level of financing provided for companies could be up to 50%.
- Venture Capital funds.

Case study of cooperation between academia and industry – Environmentally Friendly New Thermal Sewage Sludge Utilisation Technology with Phosphorus Recovery

Phosphorus is mainly used in the production of fertiliser (80-85%), detergents (12%), and as feed additives (5%). Phosphorus concentration over 100 μg P/dm^3 causes danger of eutrophication in lakes. In fresh water streams in many urban areas the phosphorus content exceeds 200 μg P/dm^3.

According to the current estimation, commercial phosphorus ore deposits will run out in the next 70-100 years. There is only one country within the EU where phosphates are produced – Finland (from the apatite deposits), and there is increasing amount of produced and stored sewage sludge in EU countries This fact increases interest in the possibilities of phosphorus recovery from municipal sewage and sewage sludge and the use of recycled phosphorus as a substitute for natural raw materials. There is a possibility of profitable phosphorus recovery from municipal sewage and sewage sludge and the use of recycled phosphorus as a substitute for natural raw materials.

The production and storage of sewage in EU countries has been increasing, which causes eutrophication and ecological problems. It is expected to reach nearly 9 million tons of dry matter sewage sludge by the end of 2005. This quantity will contain over 700,000 tons of P_2O_5 which is an equivalent of ~2.4 million tons of phosphate rock. Currently in Poland ~500,000 tons of sewage sludge dry mass per year is produced, which contains 20,000 tons of P_2O_5. This quantity should increase in next 8 years to ~800,000 tons of sludge.

Increase of interest in sewage sludge proper management resulted from public interest caused by the agricultural use of sludge. Since 1995 health dangers related to Genetically Modified Organisms (GMO), dioxine and Bovine Spongiform Encephalopathy (BSE) have attracted public attention to the problem of food production safety, with a focus on minimising human exposure to such risks.

Research and laboratory work was prepared by the Chemistry Departments of the Technical University of Cracow, in close cooperation with industrial partners.

Preliminary research was applied to phosphorus compound recovery from mixed (initial and excessive) municipal sewage sludge from Kujawy Sewage Treatment Plant near Krakow. Every day 54,000 m^3 of sewage which came from ~250,000 of inhabitants is treated. Rough sewage contains ~7 mg P/dm^3 (1.27 g of P per 1 inhabitant) and after treatment 1.1 mg P/dm^3. Every day 55 tons of hydrated sludge contained ~30% of dry mass produced and stored. From the phosphorus balance it can be estimated that that this quantity contain equivalent of 1,000 kg of 100% H_3PO_4.

After laboratory tests, the technological processes of sewage sludge drying and combustion were defined. It was found that sludge autothermal drying with the use of sludge combustion heat could be realised if the dewatered sludge contains ~30% of dry matter. The quantity of ashes after sludge combustion should be ~10% of initial sludge mass. Ashes contain 30% $FePO_4$. The combustion heat of tested sludge was 16-18 MJ/kg (i.e., if 2,000 tons of dehydrated sludge are a single load, the results should be: 600 tons of dry mass ⇨ 240 tons of ash ⇨ 35 tons of 100% H3PO4 and 160 tons of extracted ash).

The results of initial research on sewage sludge management are based on the connection between raw materials, products, waste and energy stream flow. These activities have allowed the preparation of an inventory containing different methods and techniques:

- methods for treatment for municipal sewage, modified according to cleaner production methodology regarding quantity reduction, where the use of pollution prevention efforts and in-process recycling in sewage treatment processes is predicted (as well as obtaining sludge with specified physical and chemical characteristics, best from thermal utilisation point of view),

- thermal utilisation of sewage sludge, realised with regard to the minimisation of energy consumption and production of ash which has the optimum properties in terms of phosphorus recovery,
- recovery of phosphorus compounds from ash after sewage sludge incineration and the use of produced phosphorus compound as substitute for existing natural raw materials and products.

Technology developed by researchers at the Technical University in Krakow is the subject of implementation in Tarnow's Waste Water Treatment Plant (WWTP). It is a mechanical-biological-chemical waste water treatment plant. It is one of the largest Waste Water Treatment Plants in south Poland. It cleans waste water from the city of Tarnow (124 000 citizens) and four communes located near the city. Additionally, WWTP cleans organic waste water from the big chemical factory Zaklady Azotowe (nitric chemical plants). The total capacity of WWTP is 340 000 PE (Pearson Equivalent). For today, levels of all parameters in outlet waste water are below the limit specifically set in the 91/271/EEC Directive of 21 May 1991 with reference to urban waste water treatment.

The overall objective of the project was to develop an innovative technology which satisfies environmental and economic standards for thermal sewage sludge utilization with phosphorus recovery from the burned ash, and to establish its commercial viability. This break-even technology (chemical process that makes it possible to obtain optimum results and the use of combusting heat from thermal utilisation) is designed to save natural resources of raw materials, and simultaneously protect the environment. Thermally prepared ash becomes a source of cheap and safe phosphorus natural raw material, and is expected to be a turning point in approaching the sewage sludge problem – it becomes a valuable phosphoric raw material. Introduction of this technology will reduce the quantity of sludge disposed of (an 80-90% mass reduction in sewage sludge can be gained), which is a landmark in the aspect of the major challenges for the next decade in the EU (the project of the new sludge directives). Moreover, the results of the research performed up to now at laboratory scale proves that over 90% of the phosphorus contained in sewage sludge can be recovered with sufficient quality and used as a substitute for specified phosphorus natural raw materials.

Case study of cooperation between academia and SME – new product for cosmetic surgery, body care and rehabilitation massages, baths and hair and nail care

Case Study of INVENTIA Polish Technologies Sp. z o.o. 80-310 Rotmanka, ul. Bukowa 11, woj. Pomorskie, www.inventiapt.com)

The Company is an author, owner and the only *Natural Collagen – formula 5-22** producer at the same time. Research on creating the final product lasted about one year; in that time collagen was carefully analyzed. It was biochemically, clinically and laboratory tested, which led to creating the absolutely perfect form of the active protein.

The product *Natural Collagen – formula 5-22** is obtained from noble species of river and lake fish; it is a product with gel consistency, and has transdermal characteristics. It is characterised by a stable physical and chemical structure, good absorption and is not allergenic. It does not contain dyes or aromas.

The collagen obtained from cattle skins, directed to the cosmetics industry, contains all collagen components, but is not biologically active and does not have *in vivo* repair mechanisms in place. Moreover, collagen is more easily extracted in higher yield from fish skin than from mammalian skin. Fish collagens should have a relatively low risk of possessing unknown pathogens such as bovine spongiform encephalopathy (BSE). Collagen from fish skins keeps its biological activity, because it is obtained in optimal laboratory conditions and is naturally filtered with silk fibroin. This way of obtaining collagen guarantees keeping the natural spatial structure of amino acid, which shapes up in a triple helix. Only this structure enables the collagen to penetrate through the skin's layer to the inner organs and bones.

The product (collagen) keeps its stability in temperatures between 5 to 22 °C, and because of this it surpasses the products of other companies.

Natural Collagen, transdermally delivered to merge tissue creates wire, which stops water and lipophile substances, causes epidermis growth and nipple layer, and removes the callous layer, which enables better penetration of elastin and amino acid. It enable the organism to produce the collagen by the skin's fibroblasts.

The substance is designed for cosmetic surgery, body care and rehabilitation massages, baths and hair and nail care. Consistent application of the preparation regenerates, moisturises and makes skin firmer. It decreases stretch marks, smoothes scars, regenerates skin that has been subject to burns and excessive sunbathing. The beneficial influence of collagen has been observed in treating the minor injuries (fractures, sprains, broken tendons, etc.), joint inflammation, varicose veins, paradontosis, allergies, skin changes related to acne, pityriasis (peeling off of skin) and dandruff.

Product types and their purpose:

- *Platinum collagen* is applied as a finalising agent following plastic surgery operations, in revitalising the mature complexion, following the cleaning of acne skin underneath the soothing mask, following sunbathing sessions in solarium and in order to moisturise the skin at a deep level.
- *Silver collagen* is applied in massage, with a deeply revitalising and energising action (skin dermatoses, contact allergies, following hot wax depilation and sunbathing sessions in solarium).
- *Graphite collagen* is recommended in baths designed to introduce collagen into an organism in a comprehensive way in home treatment, in sanatoria and in biological recovery centres (fitness clubs). Recommended as an energising supplement for people who have gone through great physical strain. It is also applied in order to postpone the ageing process. Also applied in hair and nail care.

Further research on product quality and the possibilities of combining it with other products, included gammalinolen acid (GLA) will be carried out. The company has observed favourable effects. In this field we intend to widen cooperation with research institutions. So far cooperation was with the Faculty of Chemistry, Nicolaus Copernicus University, Toruń, Poland, in the field of the photochemical stability of collagen from several sources (calf skin, rat tail tendon, fish skin). From the research carried out at the Faculty of Chemistry, N. Copernicus University it is known that UV irradiation is very harmful for collagen's intrinsic structure. By using a small dose of UV radiation for fish collagen, however, it is possible to get collagen with better thermal stability.

The overall objective of the collaboration between the company and academia is to develop an innovative technology with implementation of scientific results.

Summary

There is significant rise in the number of people whoh possess higher education degrees in Poland. Also the number of researchers directly employed in R&D activities increased in recent years. In the long run this should have major impact on quality and amount of innovation implemented in Poland.

Apart from that fact, current analysis of innovation issues shows a rather pessimistic picture. Unfortunately, regardless of the measuring methods, the level of innovation incentive is insufficient for the needs of the information society and knowledge-based economy. The level of innovation expenditures in relation to GDP is insufficient, and a few times smaller than the EU average. Most of the expenses are incurred by public authorities, and the corporate share should be increased significantly. Also the number of patents filed dropped significantly since the beginning of the transition period and stabilised at a low level. The proportion of high tech products exported in relation to total exports was also, on the 2-3% level, much lower than the EU average.

There are, however, a few factors which should have a positive impact on innovation incentives in Poland. First there is a growing qualified, young workforce, ready to develop and implement innovations. Research infrastructure should be built with significant help coming from structural funds. Similarly, companies will gain another vast source of financing (apart of pre-accession funds) innovations. It should also encourage entrepreneurs to organise their own research laboratories. Funds will also be available for buying patents and technologies both domestic and foreign. All of these possibilities will require tightening of the cooperation links between industry and academia.

References

CSO (2003) – Central Statistical Office, Concise Statistical Yearbook, pp. 290-291

Goliński, M. (2003), "Gospodarka i informacja – Economy and Information", in Lubacz J. (ed.), "W drodze do społeczeństwa informacyjnego … – Towards Information Society ..." p. 145

Government Strategy (2004) Government of Poland, Warsaw.

MSRIT (2003), Ministry of Scientific Research and Information Technology. State of Scientific Research in Poland.

OECD (2003) Main Science and Technology Indicators. OECD, Paris.

OECD-KBN (1999) "Proponowane procedury standardowe dla badań statystycznych w zakresie działalności badawczo-rozwojowej – Standard procedure proposal for statistics in R&D", Warszawa, pp. 18-19.

Chapter 7
Financial and Organisational Provision of Innovation Activity at Research-Based Spin-off Firms in the Transitive Economy of Belarus

ANNA I. POBOL
Institute for Economic Research, Ministry of Economics of Belarus
Slavinski St. 1/1/110
220084 Minsk
Belarus

Summary

This chapter presents some aspects related to the financing and organization of innovation activity in Belarus.

I. Introduction – Framework conditions for appearance of research-based spin-off firms in Belarus

Radical changes in economic, social and political systems after Belarus became independent in 1991 caused significant transformations in the scientific and technical (S&T) sphere. Synergetics argues that lack of stability is a fruitful soil for the new stabilising structures would be born in the open non-linear systems (CIPIN 1996). Academic science, where the knowledge serves as input and output, is inherently such a system. The transformations that have taken place in the scientific sphere of transitive economies (and are occurring till now) are acting as the mechanisms of self-regulation, study of which helps to trace the whole palette of failures (market, governmental, system, networking failures), technological trajectories, path-dependencies (technological and organisational), and lock-in situations in the process of formation of national systems of innovation (NSI) for transitional economies (Nelson 2002, Niosi 2002, Lundvall 2002).

Transformations in the system of innovation activity organisation started in the early 90s, after the period of the 'institutional vacuum'. As part of research institutes got disaffiliated from Soviet Union subordination, they acquired greater freedom in forming the strategies of their research activities and collaboration with industry. At the same time, such freedom meant a reduction of financial flows to fund research and development (R&D) activities. Breakage of the scientific and production linkages with Russia that followed the gain of independence in 1991 has redoubled the challenges which public research institutes had to face, by rapidly curtailing the financing of expenditure on R&D from the state budget. Scientific organisations, the activity of which has largely relied on budgetary financing, have encountered major difficulties. To the present day, expectations of irregularity in funding from state sources makes them search for other schemes for their R&D activities: in 2001 the proportion of own funds of organisations in expenditure on science has increased from 12.8% in 2000 to 14.3% (Ivanov et al. 2003). For this reason, the management of public research institutes tries to increase the share of economic contracts, and reorient activity of them on the final stages of innovation cycle. Those public research organisations which were already in the time of the Soviet Union more closely linked to industry, covering part of their R&D expenses through economic contracts with businesses, were better prepared for severe times of economic crisis. More than that, experience of many years of research in collaboration with industry, their more flexible thinking in organisational innovations, but most of all the treasure of contacts and precise knowledge of demand and particular manufactures in industry, with their needs and their technological problems and correlations, has served as a reliable framework within which researchers proved able to build their own technology-based business. Numerous companies have been established by researchers and with the share of 'parent' public research institutes, aiming at promotion on the market and commercialisation of scientific developments generated by researchers.

II. Methodological issues: choice of objects to study

No studies to investigate the phenomena of research-based spin-off firms (RSO), have been made in Belarus till now. The *empirical basis* of the present paper, which provides the major qualitative microlevel data, is the case studies with "opportunistic data collection" (built in many instances on the methodology of theory-building from case studies, Eisenhardt, 1989) of RSOs in the *high-tech sector of machinery and metal-processing industry (MBMP) of Belarus,* carried out (Gordienko et al. 2002), with the fate of firms interviewed being traced for the following 3 years. Choice of the MBMP sector was based on the influence which this industry has on the economy (Table 1). *Obviously, firms the activity of which is based on frontier technologies enjoy the most fertile soil in machinery and metal processing.*

Table 1: Industrial sectors where frontier industrial technologies are created, 2002

Number of frontier technologies created per sector of economy		Consisting of			Organisations creating frontier technologies
		New in a country	New abroad	Princi-pally new	
[1] Total	407	371	28	8	150
[2] Industry overall	177	167	6	4	73
[3] Machinery and metal processing (MBMP)	96	86	6	4	36
[4] Metal processing	57	49	4	4	21
[5] Share of frontier technologies created in MBMP in all frontier technologies created in industry, %	54	29.3	66.6	100	
[6] Share of frontier technologies created in MBMP in all these in economy,%	23.6	13.2	14.2	50	
[7] Public education	44	34	8	2	15
[8] Science and service of science	186	170	14	2	62

Source: calculated on data of Committee on Science and Technologies, Council of Ministers of Belarus, 2003

The significance of the machinery and metal processing sector for the economy is stressed by the fact that this sector contributes the largest share to the total industrial output in terms of production (Table 2). Nevertheless, in the structure of organisations which carry out scientific research and development industrial enterprises take a share of only 11.7%.

Table 2: Share of selected industries in total industrial output in 2003 (per cent)

Electric power	Chemical and petrochemical	Machinery and metal processing	Logging, wood-working	Light industry	Food	Other
8.1	12.8	21.3	4.7	6.4	16.5	30.2

Source: Data of the Ministry of Statistics of Belarus.

Empirical pool of objects to study

As is the case elsewhere, new companies more frequently branch off in Belarus from the large enterprises (corporate-based spin-off firms). But even in case of being connected with research, they choose the organisational form of 'scientific-production enterprises', where the share of R&D constitutes from 10 to 15%, and thus do not contribute much to the development of a national innovation system. That is why in current analyses attention will be concentrated on *research-based spin-offs from public research organisations* and *from universities*.

In Belarus, the term 'research-based spin-off' has not yet been introduced into policy practice, neither are there any statistics for this sector of innovation economy; thus, it appears difficult even *quantitatively to estimate* the two groups. The *general borders* of the pool to which RSOs belong are those of small innovative entrepreneurship in general. The definition of these in Belarus is rather limited, but thanks to the application of strict criteria a clearer picture can be provided. Official categorisation as "innovative" strictly demands that an organisation should perform R&D, with the size being defined as "small" according to a sector-based scale (Table 3). There were 317 enterprises (1.2% of small entrepreneurship) in Belarus in 2002 which met the requirements of such a definition. Small innovative entrepreneurship thus represents the upper theoretically possible border for the pool of objects to study.

Table 3: 'Staff on the payroll' criteria of firm attribution as 'small entrepreneurship' subject'

Activity in the sector	Number of employees
In industry and transporting	Up to 100
In agriculture, including farming, and in scientific and technical sphere	Up to 60
In construction and wholesale commerce	Up to 50
In retail trading and household services	Up to 30
In other spheres of the nonmaterial sphere	Up to 25

Source: Law of Republic of Belarus "About the state support of small entrepreneurship in Republic of Belarus" of 16.10.1996 № 685-XIII

Small business, being generally a very narrow sector of Belarusian economy, is less and less engaged in science-intensive activity. As stated in "Concept of State Support and Development of Small Entrepreneurship by Council of Ministers of Belarus for 2002-2005", only 27.8 thousand legal entities – subjects of small entrepreneurship – and some 180 thousand individual entrepreneurs were operating by early 2002. There are only 2.7 organisations with the status of small entrepreneurship per thousand inhabitants in Belarus, whereas in Russia this figure equals 6, and in the European Union 30 (in the USA, 74). One of the officially recognised trends is the preferred orientation of small business towards trading and agency operations – 42.8 % of all entrepreneurs concentrate on the sphere of trade and catering consumption, whereas in industry only 21.2 % of them do. (Entrepreneurs of the science-intensive sphere explain this phenomena in terms of "*slow*" (in the R&D sphere) and "*quick*" (in trade) cash flows and the general challenging economic situation; but also the skills requirements are different for these kinds of entrepreneurship).The share of small business in R&D diminishes every year. It had dropped from 601 enterprises in 1997 to 317 in 2002, and constitutes a proportion of only 1.2% in the total number of small entrepreneurship businesses (Table 4).

Table 4: Small innovative entrepreneurship

	1997	**1998**	**1999**	**2000**	**2001**	**2002**
Number of small innovative enterprises (SIE)	601	537	503	412	348	317
Share of SIE in total number of small enterprises	2.8	2.2	1.9	1.6	1.4	1.2
Share of employees at SIEs in total number of employees at small enterprises	2.7	2.4	1.7	1.1	1.0	0.8

Source: Ivanov et al., 2003

Only one tenth of the SIEs set up over the last 10 years in Belarus have survived, and a only few of them were capable to develop their activity up to the level of mature small innovative enterprises, keeping the science-intensive activity as their major one.

Clearly, entrepreneurship in the academic sphere, or sector of enterprises created by former employees of public research institutes, which are the subject of investigation in a given paper, is even more restricted. Trying to quantify the division between *university-founded spin-offs* and *spin-offs from large public research institutes (LPRI)* implies difficulties. Here it is worth mentioning that research in Belarus is conducted in specialised organisations with an affiliation either to the National Academy of Sciences, or to various Ministries (Ministry of Industry, of Economy, Education, of Statistics, etc.), or again, to large industrial enterprises (concerns –Belneftechim, Scientific and Production Concern of Powder Metallurgy, Belmashpribor, Belbiofarm, etc.). To a great extent the scientific and technological potential of Belarus is concentrated on its universities; a number of developments in this quarter have started to get successfully commercialised. For the organisational support of university R&Ds, the Belarusian Interuniversity Centre for Marketing of Scientific and Technological Developments operates. It unites in a single network the 21 universities from all over the Belarus, and runs a portal, where the profiles of 601 ready-to-market developments of Belarusian universities may be found – the objective basis for the creation of a minimum of 601 university spin-offs around them. Still, no exact statistics about the number of firms already created is available. As for spin-offs from LPRIs, technological profiles of those launched by National Academy of Sciences are published yearly. Thus, as the quantitatively determined *bottom boundary* to estimate the extension of RSO sector in Belarus one has to accept the figure of 30 enterprises, which have been launched in a system of the National Academy of Sciences of Belarus (NANB), and are operating in the form of small innovative, introductory, and other entrepreneurial structures.

III. Conditions frameworking financial and organisational aspects of RSO activity

Organisation and funding opportunities of innovation activity for RSOs are preconditioned by the systematic condition and specifics of both spheres of economy (science and industry), at the crossroad of which this research-based entrepreneurship stands, and the gap between which it aims to fill in. The place in the economic system and *operational conditions of small entrepreneurship* require closer analyses, as far as RSO belong to this sector, and may apply for corresponding support. Also the position of the *state R&D sector* (academic, university, and branch science) and experienced attitude to it are influencing the dynamics of RSO sector, because these firms originate from the scientific world, and have mostly close cooperation with public research organisations. Above, some RSOs (with a status of state ownership) apply the relevant taxation schemes. Lastly, a factor of severe consequence for the functioning of science-intensive business is the financial and technological picture for its *industrial customers* – most significant of which are the large factories.

1. Financial and organisational support of small businesses

Financial and organisational support

Direct and indirect state support of small business is carried out in practice in a framework of measures of the state and regional programmes for state support of small entrepreneurship. *Financial* assistance is provided to small entrepreneurship for the accomplishment of their investment projects, which are focused on the organisation of new and extension of current productions, on the manufacture of competitive, export and import-substituting products, and also for projects which contribute to the creation of new working places. The *mechanism* of such support is the partial compensation of expenditures for bank credit interests. Above that, partially the expenditures on payout of leasing payments on contracts of financial rent (leasing) are compensated (in part of bank interests for usage by leaser of borrowed funds and expenditure for participation of small entrepreneurship in exhibitions and fairs). Small enterprises are also supported *organisationally*: through provision of information, consulting services, education, through the network of infrastructure organisations to support small business that exists in Belarus, which as of 1 November 2003 included 43 centres and 7 incubators. In 2002, 43,588 persons applied for such services, and 27,822 for a period of 9 months in 2003; 592 educational courses and seminars were conducted in 2002, where 11,065 persons involved. 123 subjects of small entrepreneurship have been situated in incubators of small business in 2002, with the total premises rented coming to 5316.7 square metres. Nevertheless, the application rate seems to be rather shrunk when we compare it with the number of individual entrepreneurs, which is an average of 190 thousand persons, and the 26,849 small enterprises, 29,044 of them in private ownership form, with the number of staff employed coming to 255,622.

Credit funding

Difficulties in access to bank credits for small entrepreneurship in Belarus are serious obstacles to their becoming more competitiveness. Though 70% of small enterprises suffer financial difficulties, only 27.9% resort to bank credits: of these 14.1% for financing their investments, and 13.8% for crediting their current assets (Metolit 2003). Instead, small business uses the credits from individuals and partnering small enterprises. Among the factors which restrain the opportunities for small enterprises to take bank credit are the very rigid requirements to SMEs from banks in what concerns security for a loan (including guarantees), and the lack of system of guarantees and leasing mechanisms. Reasons for this are the lack of solid credit history of young enterprises, the higher than usual production risks and low capital mobility, and the structure of assets of a low liquidity: having the valuable human capital (intangible assets), they rarely possess real estate, which could serve as caution or guarantee for a bank credit. Some additional factors that result in small enterprises preferring to buy credit services by individuals are the lower transaction costs in this scheme of fund-raising, the time spent on filing the credit, the possibility of a longer period of credit use, the possibility of receiving the cash and friendly relationships with creditors. Lastly, bank credits are very expensive – interest rate on bank loans is twice as large as the interest rate on loans from individuals (40-60% as opposed to 20-30% of private loaners).

2. Financial and organisational support of science

Funding opportunities

Financial resources of science are officially accounted for in Belarus as the internal (capital plus current) expenses on R&D, which are executed by reporting organisations, and the works of co-executors. In 2002 internal expenditures on R&D constituted 162.3 billion Belarusian rubles (BRB) (which comes to 0.038% of the state budget), of which capital expenditure came to 6.3 billion BRB. Giving consideration to the ratio of expenditures on R&D and gross domestic product (GDP), it should be noted that the science-intensity of the GDP in Belarus amounted to 0.73% in 2002 (see also Table 5). Even the application of the method for assessing the science-intensity level used in the Commonwealth of Independent States (CIS) (Expenditures on research + developments + S&T services)/GDP in opposition to (Expenditures on R&D)/GNP), commonly employed in the EU, shows the level of science intensity to be less than 1%, which is a critical one even for the survival of science.

Table 5: Volume of work executed by scientific organisations, as a ratio to gross domestic product (GDP) in selected countries of CIS.

	1990	1995	1996	1997	1998	1999	2000	2001
Belarus	2.11	0.93	0.87	0.81	0.78	1.04	0.81	0.81
Ukraine	2.33	1.34	1.38	1.36	1.22	1.24	1.14	1.1
Russia	2.98	0.81	0.98	1.06	1.1	1.22	1.28	1.4

Source: Ivanov et al. 2003

Most research and development projects with direct practical application to industry are carried out and realised in a framework of the 'State scientific and technical programmes' (SSTP), innovative projects, branch and regional scientific and technical programmes. In recent years projects close to completion have been favoured, with the aim of rapid commercialisation. For such projects, 'organisations-executors' periodically collaborate with companies which have spun-off from them, because of the greater flexibility and concrete experience of solving the practical problems of these. The scheme of collaboration in SSTP has as a prerequisite the compulsory participation covering a minimum 50% of R&D expenditure of organisations – customers of executed R&D. Many RSOs help their parent institutes by finding the enterprises which would agree to the SSTP collaboration scheme (overcoming the usual scepticism of industrial executives about the duration of R&D in academic science).

Taxation

The level of taxation is pointed at by innovating entrepreneurs as being among the most burning problems for the development of entrepreneurship. The *tax burden* in the branch of science and service of science constitutes in 2002 21%, being accounted for as the share of taxes and payments in sales receipts, which is, as also in all previous years of reformation, higher, as the tax burden in industry and economy as whole. This situation is caused not lastly by the high share of wages in the structure of expenses on scientific and technical production. The share of expenditure on wages and social needs in the structure of expenses constitutes 39.5% in the branch of science and service of science, in comparison with 22.0% generally in the economy and 18% in industry (Ivanov et al. 2003). This causes large payments of taxes which are calculated in direct ratio from the gross payroll size.

According to current legislation, in Belarus several kinds of tax benefits are applied for the stimulation of R&D and innovations, independently of property form and the sources of financing of economic subjects: the appropriate expenditures, when being accounted in costs of production, are exempt from VAT and profit tax. Nevertheless, these preferences do not render significant assistance in lowering the tax burden for science-intensive activities. In 2002, all the kinds of R&D that were financed from budget were largely exempt from taxes. But the enterprises and organisations of branch science, the activity of which is mostly financed from non-budgetary sources, and certainly the private science-intensive business have been lacking benefits in terms of profit taxes, VAT, etc. Also, there are no taxation benefits for the innovation stage of mastering the innovations in production, which require the most investments and cause the most strained situation with cost-performance ratio.

3. Financial mode of RSO customers

Customers of innovative technologies belong to both private and public spheres; these are the small enterprises of private business, middle-sized enterprises (of mostly mixed ownership), but mostly the large industrial enterprises and public research sphere.

Obviously, SMEs of various sectors of the economy finance introduction of their new technologies from sources with different structure. Nevertheless, the average of them may be taken as representative: this indicates that most rely on their own sources (78%). State and regional budgets as sources for the financing of innovations of SMEs (those owned by the state) have a share of only 0.5 and 0.1%, and foreign investments' share constitutes only 2.3% of all expenditure on technological innovations.

Table 6: Factors prohibiting introduction of scientific developments by SMEs,% of responses

27	17	13	12	10	5
Lack of financial and material possibilities	Absence of correspondding tools of state support	Absence of good, intelligible information about technologies	Rather difficult state support usage procedure description	Trust to foreign technologies	Absence of technology transfer centres advertisement

Source: Metolit 2003

Large enterprises with state and mixed forms of ownership (joint ventures with partial state involvement) provide the largest contribution to the economic system of Belarus, and form the major share of demand on the technological market. The enterprises by themselves have an outdated technological base (with 60-70% of deterioration rate average in industry), and require the critical restructuring and introduction of new technologies and equipment, which would allow them to develop new products. But because of the economic crisis convoying the transitional period of economy, their demand on innovations becomes insolvent, and they face large difficulties in financing innovations. Still, in the light of data on monitoring by programme of industrial complex development in Belarus for 1998-2015, 308 new most important kinds of industrial products and technologies were mastered in 1998-2000. Large industrial enterprises themselves participate in the creation of innovations. 38% of technological innovations introduced in 176 enterprises from 9 leading sectors of industry with a size of more than 100 employees in 2000 were carried out through the co-operation of industrial enterprises with other organisations (Ivanov et al. 2003).

There exists an internal contradiction in the economic system of Belarus between the necessity of renewing the technological structure of the economy, and connected with it, low labour productivity, which restricts the solvent demand of population and profitability of production as a base of technological upgrade, thus creating a *lock-in* situation. One possible solution could be to attract external sources of investments and international cooperation in R&D projects. Objectively estimating the current situation, one should mention the comparatively low integration of enterprises into world scientific and technological space: enterprises cooperate in R&D mostly with domestic organisations (77% of the total number of collaborative projects) and Russia (16.6%). Of all the collaborative projects of industrial enterprises (253), only 9 have been executed with organisations from other CIS countries, and the technological regime used was that of technological export – the number of transmitted technologies has been significantly higher as this of acquired. Thus, such a technological orientation of enterprises of Belarusian industry should be taken into account when considering the development of technology-intensive business.

Large industrial enterprises appear to be currently one of the most vulnerable elements in the innovation system of Belarus, having difficulties not only in funding the creation of the technological innovations they need, but also in the introduction and adoption of them in the production process. The major and most significant factor, blocking up innovations, is that enterprises are not only lacking the financial possibilities to fund their own modernisation and restructuring on the base on development and introduction of innovations, but do not possess sufficient incentives to innovate.

IV. Schemes of organising the activity of research-based spin-offs

1. Operating on market

Niche on market. Customers

All the firms have now their basic R&D-intensive activity as the only one, though some of them also had to execute external to core innovation projects for the accumulation of capital at initial stage. Now they outsource all the jobs which do not directly belong to the technological process. Changes in manufactured products usually result from the further development of ideas underlying the basic technologies of each company, and are modified flexibly according to the special requirements of customers. Main markets of researching entrepreneurs are the CIS countries; till now only a few of them export to the west (cooperation with the Far East is emphasised). Still, Belarusian developments definitely 'outspeed' western competitors in terms of the price and reliability of equipment, some also in technological characteristics, practically displacing European and American competitors from the CIS market. In some technologies, good potential exists for Belarus to occupy this market niche on the world market practically completely. The *competitive advantage* of Belarusian RSOs consist "in advanced knowledge of technology, research and developments, scientific background for processes". Russia provides a good marketing area. It is even easier to organise cooperation in Russia than in Belarus (because of the insolvency of Belarusian enterprises-customers and interest of Russian customers in further R&Ds). Most successful Belarusian RSOs have Russian market as the target one, with the Belarusian market being confined to a minimum.

In spite of restricted financial opportunities, it is small enterprises which turn out to be most responsible customers. Simultaneously, large enterprises give the largest and most serious orders, but delay payments of debts to small enterprises for months. The small company can however "do nothing to make the state giant repay its debt". Similarly, it turns out to be easier to collaborate with non-state enterprises, because in this case the projects are paid more accurately. By state enterprises payout can be late for up to half a year, which is irrelevant for RSOs when the period of order execution lasts 3-4 days. Because of differing necessities and financial abilities of customers, RSO managers have introduced *alternative forms of cooperation*: delivery of ready details to customers; delivery of raw material, semi-finished items for further production on the customer's premises; licence sales; adjustment of manufacture with delivery of the necessary equipment to the factory; joint manufacture of production (with its patenting) for internal consumption in Belarus and for external countries, etc. Most contracts are concluded on a *one-off basis. Contracts on a more continuous basis* are rare – such contracts are affordable only to customers with stable products (which will be guaranteed reclaimed, or products with standard characteristics, that are applicable even in case of final product modification), but they are also concluded.

Diverging tendencies are detectable a propos the *stage* at which the customers are ready to accept innovations of RSO: if Belarusian consumers prefer to buy the final products (services of RSOs executed on their installations, for example, superhard tools for cutting the superhard materials, or vacuum coating, etc.), eastern customers (South Korea, China), strive to buy the technologies, and western ones (and Japan) to get the experts themselves. Outstanding preferences of Belarusian consumer-enterprises, which all without rare exception require renovation, is explained by the fact that tje introduction of innovative technologies causes the need of modification of equipment, significant reduction of personnel of the plant, and investments in qualifications of these who work on new installations; these very reasons hamper introduction of innovative technologies, because factories are not able to finance the total restructuring and layoff of employees.

Market development: barriers to widening of market

The commercial success of RSOs is relative. Profitability levels range from 3-4% to 15-20%. Most provide their founders with a modicum of existence, with the wages higher than at parent research institutes; some do quite well, and there are only 3 or 4 which have managed to achieve very successful commercial results. The reasons are – 1) the high technological level of products and services, their science-intensity needing a large proportion of expenditure on research projects before offering the products to the market and during the promotion of tools; 2) the absence of demand on production in Belarus, since the common policy is the orientation on decision of social, instead of industrial problems.

Two polar cases are of interest when considering the *attitude of RSOs to development* of their market. In one case, managers of RSOs do not expect to expand their activity. Even if the major customers of RSOs are Russian enterprises, which helps to overcome the barrier of customers' insolvency, there remains the barrier of progressive taxation, "which by itself can lead to the extinction of the whole small innovative business in Belarus". But also, there exists only *low motivation* for Belarusian RSOs to expand their activity on European markets – there is no sense in this when the Russian market provides such a volume of demand that the capacities of the enterprise are hardly sufficient to cover it. This market is 'guaranteed' for some RSOs, which have no competitors in the whole territory of CIS, and entrance barriers to their technological sphere are rigid because of the qualifications and expensive equipment required. Experts at such a professional level do not long stay unknown among colleagues. Certainly, there is no talk about organising the subsidiary firm or spin-off company of the second generation in order to keep the monopolistic position on market attributed to them by unique technologies.

On the second pole is the case of a highly promising technology without analogues in the world for its technological advances. The core technology (nanomaterials) will allow development of the wide cluster of further high-tech spheres with their own technologies and specific productions in many different industries (automobile, medicine, chemical, etc.). The possibilities of new technologies are so vast, and the market is so unprepared to accept this spectrum and adapt it in production, that one of the major concerns of the innovators has been how to attract competitors, who would "heat" the market by their simplified versions of technological processes.

External support:
national and regional programmes and establishments for SME assistance

88% of small enterprises have never received state support, 6% have received it, the others receive it irregularly. The reasoning of RSO companies which have never tried to apply for any kind of state support is based first of all on the uncertainty of managers as to whether they will be financially solvent in the near future and will have the opportunity to pay the credit back, even if they sell the equipment of the company. The 'second worst' reason is that "all the newly established organisations" aimed at providing certain privileges to innovative companies, with the privileges being as a matter of fact very small, request such a quantity of paperwork and reporting for them, that it is easier to manage without these privileges. "Accounts are measured no more in number of copies, but in kilograms of paper". Moreover, the activity of the company after a successful application for support is immediately subjected to interference by plenty of controlling authorities, making the normal operating of the firm impossible.

The experience of those entrepreneurs who have received state support ranges from definitely unsatisfactory (for reasons of small support budget, long procedure of documentation assignment and insufficient qualification of employees as for consulting-supporting activity) to very positive, because even this support allows the execution of large projects which otherwise could never been organised. State support is certainly considered useful. However, the basic conditions for the normal development of innovative activity, desired by innovating researchers, are market economy principles, market-based price formation and a normal investment climate. "I do not need any infrastructure – had I money, I would then easily organise everything by myself. People who are external to the company do not understand what is actually necessary for me – which information, what materials for production ..." – one of respondents says. "The best the state could do is to let the enterprises work peacefully".

2. Resources

Sources of funding the innovation activity

The *initial capital* in most cases consisted of own savings of researchers. Problems with the financing of manufacture and further R&D are crucial. Funding of activity is raised through concluding *contracts on particular jobs with customers* as well as *R&D activities under the state programmes at country and regional level.* Developments are carried out within the payments on particular orders. There are no means for conducting independent developments, nor are external sources used. A representative example of financial schemes of RSOs operating by the state research-based enterprise: nearly 50% of expenditure is financed through participation in state scientific and technical programmes (SSTP) and separate projects. About three percent constitutes funding from the state budget, and the remainder comes from individual contracts with customers. Private firms are not allowed to participate in SSTP, so they have to rely completely on contracts. Actually now all activities are carried out on the self-repayment basis. All developments are financed from the own means of the company earned from contracts. Nearly 30 per cent of earnings is allocated for salary payments, the rest (which depreciates quickly because of inflation) is spent to supplement the working capital for conducting the manufacture projects, and to finance new developments. Lack of financial assets often makes it impossible to realise the fullness of ideas, or even plan current activities. Lack of finances and material support is generally the most burning factor preventing SMEs from introducing scientific developments (38% of responses).

To *eliminate the consequences of financial insolvency of customers* at least partially, the companies use now the scheme of 30% of advanced payment instead of 100% at the beginning of its activity. However, even in this case the customers whose activities are financed from the state budget practically always have difficulties with timely payment of bills. They give the largest orders, and they are also the most unpredictable partners. For example, payments for executing the jobs are received only few days before the term when the order should be already fulfilled. Meanwhile, economic conditions at the enterprise itself do not make it possible to produce the goods in advance ('work for warehouse'). First, practically no circulating assets are available – all works on orders can be carried out only after the advance payment has been made. Second, each customer has his sophisticated requirements to the style of products, special characteristics (degrees of protection) they provide, etc. Innovators have to work *for only the customers' specific requirements*, and try to raise money for developments from customers' payments of their concrete projects. Until recently, rapid depreciation of money in Belarusian unstable economic conditions made it necessary to spend funds immediately. The funds from projects with payment in advance were directed to financing the expenses of currently running projects.

It is not possible for an ordinary RSO company to raise *bank credit* for carrying out R&D or broadening its activity. There are no investment credits available, and even those which can be received imply a very high interest rate. To receive financing, either informal assistance is necessary, or a guarantee. Banks "do not need our old broken vacuum installations" as a guarantee, but the receipt of the credit is connected with burdensome bureaucratic formalities and requires a load of papers to be filled in. Innovators do not want to get involved in this procedure – one needs to spend about a month for registration of the documentation required, and then it can quite possibly appear that because of one insignificant reason the money will not be allocated to a firm. Procedure of financing reception is too long and overcomplicated. "Why should I totally take the credit, whether from a bank or an innovation foundation? If I cannot sell the installations designed, I cannot pay the credit back" – says a respondent.

There has been only one enterprise among those studied that ventured to take *bank credit* for the creation of new manufacture; now payment of this credit back is among the most serious problems of the enterprise. At a preparatory stage, the own funds and resources were spent on technological development, as well as extra funds of bank for starting of manufacture, but at the moment when the stage of the first receipts from sales should have started, the market appeared to be unprepared for accepting the innovative products. The bank credit was to be allocated only for material assets – buildings, constructions, equipment, etc., so that they could serve as a pledge of return of funds; expenditure could however not involve working capital for the beginning of activity, nor expenditure on scientific projects, certification, advertising or preparation of market.

The bank credits for wage payment (a financial mechanism designed to provide a framework for the execution of a law dating back to 2001, according to which the payout of wages should not be delayed) have however proved to be of real use; the enterprise takes them regularly. They are not expensive, and allow having no problems with delay of salary payment.

There are no realistic prospects of financing from external investors, since "there are no funds in Belarus, and investors abroad do not want to allocate their funds here". Attempts to raise the interest of *foreign investors* have been made, but with very little success. Threats frightening the foreign investors are first of all the investment climate, and the absence of prospects for the market-oriented development of the Belarusian economy. Attempts to attract foreign investments for creation or manufacture of own expensive raw materials have been made, but this idea has failed because of the bureaucracy of Belarusian officials.

Premises, equipment, components, raw materials and supplies

Small entrepreneurship is mostly developed in the capital city Minsk, where 52.6% of all small business legal entities and 27% of individual entrepreneurs are situated. In small towns and villages it is very little developed. This tendency is followed up by distribution of scientific and technological resources – major capacities and researchers are concentrated in the capital city or in main region cities, at universities or science-intensive large industrial enterprises. Case studies show that regional distribution of research-based spin-offs keeps to this tendency: as far as science-intensive production of high-tech sphere in machinery and metal processing requires powerful sources of energy, developed infrastructure, equipped premises, firms are usually situated in direct proximity to them.

Most really do "quasi" (Radosevich, 2002) leave from parent research institutes, bearing an official status of separated, but renting the *office and production areas*. Some, less dependent on unique and very expensive set-ups, and more 'rich' to organise their own production, buy or rent some larger manufacturing areas with developed manufacturing infrastructure (for example, in the industrial zone of heat power plant, or institute for nuclear energy) and equip them. There are also cases, when production is organised close to (potential) customers – one firm rents areas at the gear plant, as far as technology of them is finely adjusted to hardening these gears on a large scale. Even though establishment of continuous cooperation between the plant that lets out the premises and the firm renting them entails tough organisational problems, innovators reckon on success as being situated there. Frequently problems arise with the payment of rent, because of the specifics of technology-based enterprises – the high share of research works, which do not give reimbursement immediately.

As is the case with the premises, *equipment* also is often rented from the institute where the researchers come from, because of its uniqueness. Also, parts of the components for equipment may be rented by the parent organisation (or founder) of the company. For technologies based on common ideas, the relevant equipment may be purchased and the company's own equipment installed subsequently. Only few firms buy equipment abroad (mostly personal computers). For importing technologies there are not enough financial assets available in the companies' budget. Moreover, homegrown technologies are considered to be better. *Completing details* of foreign production for the equipment developed by the company are probably more reliable, but they are much more expensive. In Belarus and in Russia, completing details for the equipment can be bought "for nothing" at the factories which become bankrupt and sell off their manufacturing equipment.

Relations with former (parent institute) management in some cases are rather strained; one of the reasons for establishing one's own firm was the absence of opportunity of developing some ideas because of difficult relationships with management. In such cases no equipment designed by employees of laboratory specially for conducting the R&Ds was passed to the new enterprise, and they had to find other solutions – such as the development and construction of installations (and then with better characteristics and capacities than the trial version left by the parent organisation) by themselves, or, when this is not necessary, purchasing equipment from other organisations, possibly modifying it in keeping with the researchers' needs, some from the military industry. Also the enterprises which have worked out technologies never mastered before (some even on the world level) develop equipment and facilities for assembling it simultaneously on their own. In this way innovators are able to develop innovative installations based on their technology for external organisations and consumers. When development of equipment occurs on a commercial basis, technical project and outline sketches are worked out and passed, according to which the customer constructs the necessary equipment himself.

In what concerns *raw material supplies*, there is much variety with reference to its accessibility, depending on the specifics of the technological process. Some small science-intensive firms experience difficulties with (raw) material delivery, because of the small quantity of it required (for experimental production, for example) – in Belarus, it is impossible to find a factory prepared to sell a small quantity of metal. It turns out to be even more convenient to buy metal in South Korea, because in Belarusian supply chains of such compounds up to 15-20 persons stand on the path from large wholesale up to retail. The others have practically no problems with purchase of materials: "one needs only money", because "it is possible to buy almost everything needed for the manufacturing process in Russia". For some processes, the *waste material* of heavy industry (for instance, metal rasping) serves as the raw material for high-tech new materials. Some (production of nanodiamonds) apply the materials of *military productions* (explosives). But sometimes problems with equipment, materials and components are associated with the character of this production – for the high-tech materials required can only be available from some few oligopoly manufacturers, which set up world prices (aramide fabrics). Some high-quality installations required for production are manufactured abroad, but their prices are at least ten times higher than prices of the Russian equipment.

Information resources, ICT

Difficulties in access to reliant and intelligible information prohibit the innovation activity of 18% of respondents. Absence or lack of information usually blocks the decisive actions in a definite direction. More than any other kind, information about prices is lacking, as well as about the entrepreneurial trainings and seminars held. Only 8% of respondents indicated the lack of information about equipment and technologies, which can be on the one hand interpreted as the low level of focus of small entrepreneurship on innovation activity, and on the other as a competence of those who need this information (innovating entrepreneurs) in sources and carriers of information like this.

Usage of the Internet as the mean to disseminate information about a company's own innovations is found to be very ineffective by practically all Belarusian RSOs. The major factor that limits the efficiency of information and communication technologies in the innovation process in Belarus is the low involvement of customers (mainly the Belarusian and Russian enterprises) in the commercial use of ICT. The impression of respondents about the reason for failure of advertising through the Internet is that it is of no use just to wait when one notices the advertisement of company's opportunities, especially abroad – one needs to act with more initiative and pay personal visits. Though the information of companies is placed on the websites of Belfact (Belarusian information agency), Europage, of Belarusian commercial and industrial chamber, in electronic catalogues of Belarus and Russia, the greatest feedback still is brought by personal contacts and links in scientific and factory circles. Information and communication technologies still make a very small contribution to the establishment of business contacts, and their insufficient use "is far from being among the most important restrictions for company's development".

Computer resources are mostly used as a part of technological process itself – for calculations, finishing of technology and final improvement of process. Technological opportunities of installations are considerably extended by equipping the installations with modern computer utilities, which allows process optimization.

Attraction of personnel for development of innovations, problems with staff

An important base for the attraction of good experts is the existence of well-established interconnections and contacts from the time of postgraduate study of the director and with colleagues at the institute. These, as well as cooperation with external organisations and businessmen and previous scientific and industrial activity, help to attract qualified staff. Many workers are very responsible and spend 18 hours per day on the experimental site. But with some workers there exist problems in connection with their irresponsibility and unreliability, mainly with workers who have earlier been employed at the state enterprises. Nowadays the workers with average abilities and less initiative have stayed put at factories and plants – the best have launched private business or changed their jobs for more action and perspective. Workers are hired on the base of professional qualities, and then their abilities in concrete activity are estimated. If a person does not meet the requirements, he is dismissed. Change of structure of workers occurs most often because of non-uniformity of intensity of orders. That is why successful RSOs pay high wages to its workers – hence they "can afford to choose the workers, even testing the colour of their eyes" – in order to have no problems with staff. Tough discipline is maintained among ordinary workers – for the smaller violation of labour discipline (alcohol in the workplace) they are fired.

The other general problem with staff consists in "interbelarusian brain drain", when other companies offering more favourable financial conditions for work entice the young experts prepared by the enterprise. Thus on the base of former employees of one RSO, two 'second generation' private companies in Minsk were set up, employing the parent technology, and are commercially more effective. Actually, they use free-of-charge the technologies developed by the parent RSO, and steal their customers.

Needs for training and skills in connection with innovations

The labour-intensive and responsible activity of science-intensive production demands *high qualification and advanced skills in the workers. All employees must be qualified experts.* As the issues the company deals with are complex, external experts are also involved in innovative activity, in addition to the company's own workers. Temporary scientific groups are gathered, experts from both industry (from factories), and from higher educational establishments are involved. The *young staff* are involved in postgraduate courses, they are trained, qualified experts are prepared from them within 3-4 years with the aim of their continuing work on the enterprise. *Professional education and learning-by-doing* are the common practices employed in companies to prepare the necessary experts. Experts, researchers and developers of innovations themselves constantly work to enhance their skills. Still, RSO managers are definitely lacking in *managerial competencies, knowledge and experience of economic operations*. Tackling the general economic difficulties is complicated, because the primary experience of managers and workers is gained in the sphere of scientific research, and they are still focused on solving scientific issues rather then economic ones.

Also, a serious lack of innovative competence in operation of the complex equipment is admitted to be common in the *staff of customers*. Qualifications of workers at factories where innovative equipment is introduced, as well as the level of responsibility, prove to be extremely unsatisfactory, according to respondents. It is quite natural that they are not able to use all the useful properties of the multipurpose complex installations (vacuum, laser engineering, advanced plasma technologies). All the RSOs which introduce equipment take the low innovative competences of customer's workers into account: alongside with works on introduction of the laser equipment at the enterprises, they provide *training and consulting of enterprise-consumers' staff* after the equipment is introduced, during the one year of a warranty period. Also, the installations by themselves which are produced by domestic RSOs are manufactured in full awareness of the use regime they will have to withstand: they possess much greater reliability, are more intelligible and easy to operate.

Statistics shows that the number of experts engaged in R&D in industrial enterprises of Belarus is tending to fall. In 2001, in the largest enterprises of industry, there were only 0.66 researchers and engineers working per thousand employees (Ivanov et al. 2003). The corresponding figure for the EU is 2.5 (1997). This trend can potentially cause not only an inability to create the innovation products, but also the inability to perceive and adapt in production the technologies which have been created by others.

3. Organisation of activity

Succession of knowledge resources

Founders and researching staff. Most *managers* of RSOs have also been managers in laboratories of public research institutes, or frontier subject lecturers at higher education establishments. In some cases, for instance when technology finely fits the market structure (from the point of view not of request for technologies, but of solvent demand), they fully switch over to their own business; but such technologies are rare, and most developers prefer to *continue their employment at parent research institutes*. This helps them to keep in touch with the latest scientific developments and to be integrated into the community of scientists with a view to exchanging opinions and finding mutual solutions to technological issues. The *core* of the firms consists of *employee-researchers*. Most have been engaged with issues that constitute the special field of their RSO now professionally during their earlier career. As a rule the team members have a doctorate, and are recognized experts in their scientific field. "That is what we can do very well" – one director says. The process of team-building often occurs on the basis of former cooperation (one laboratory) of founders.

According to statistics, 68% of managers of small enterprises have achieved a technical qualification, 14% humanitarian, 6% in physics and mathematics, 4% medical and 8% other. In SMEs 92% of managers have a higher education qualification, 1% comprehensive and 4% a specialised (technical) one. When we compare the level of managers' education among the SMEs in general, and RSOs, only 4% of managers have a scientific degree in

SMEs, whereas in RSOs this figure is *close to 90%*, and even 75% of non-managers have a higher education degree. Age trends coincide: 46% of SMEs managers are between 30-40, 31% between 40 and50, 14% over 50, and only 10% belong to the age category under 30 years old.

Most founders have dealt with processes which lay at the base of innovation since the 60s or 70s (at their initial stage of development at that time). Most have more than 3 patents and numerous licences, some have awards and diplomas for their innovative activity and for technologies they have developed. Being among few who developed the core technological innovation in the Soviet period or stood by its springhead, they became an ideological source for the creation of innovations. At present, developers provide both theoretical advances and the practical embodiment of ideas in technological production, and organise the commercial side of the activity.

Employment schemes and functional divisions. Practically all RSOs strive to have the slim organisational structure, and become ever 'fitter' over time. The average number of staff constantly employed at scientific innovation enterprise has decreased from 13 persons in 1995 to 8 in 2001. Only employees constituting the core of the firm are employed at the enterprise on the contract base, constantly; they organise all the activities at the company and all the innovation (R&D) works. Even in larger spin-off firms, management combines the functions (the CEO is simultaneously the design manager, the technical director is production manager, the deputy director is also the quality manager; even ordinary workers are selected on condition that they can combine the functions, for example, of lathe operator and borer). The rest of the staff is involved on the timely base, for concrete (ordinary) jobs of manufacture. For the direct execution of jobs, work on manufacturing devices, individual businessmen are involved. All of them collaborate with the RSO company on a subcontract basis – thus, the opportunity offers of lowering overhead expenses. This facilitates reduction of the excessive tax burden at the enterprise: in case the worker is employed at manufacturing works on a non-constant basis, for the one-off executing of jobs, only income tax is charged on them.

In case of joint ventures with numerous shareholders, these also do participate in various forms in the activity of the enterprise, but only few of them are employed at it directly – for instance the Director and Deputy Director (earlier the shareholders were more willing to participate in the work of the enterprise personally, but severe conditions of activity have deprived the majority of them of incentives to do this). Approximately half of RSO employees deal simultaneously with scientific research, and development of new technologies, designing, creation and sale of the equipment, – i.e. the whole complex of activities from the conception of the idea through to its actual implementation as manufacturing technology. For performance of all work which does not relate to this activity, additional workers are involved (for preparation of materials, sharpening, etc.) (outsourcing). *The more successful the introduction of technologies into industry (commercial success of innovation), the more does its structure shift to employing more ordinary workers on one researcher.*

Internal R&Ds and technological cooperation, contacts and relationships

The result of establishing the company's own production has been not only the possibility of developing the innovative technologies and equipment from the trial stage at which they have mostly been before launching the firm up to the 'logical perfection' of world level, but also the possibility of conducting research on the regular base. At many RSO companies, R&D activity has become more intensive than during former work of researchers at public research institutes and even more so than in universities. Each contract gives the opportunity to modify equipment if necessary, and to work out new technological regimes (by working with different alloys, processing metals with different characteristics). "Working with customers means the requirement that the technology be 100% reliable – it is not a report about the performed research work, which will be shelved at once".

The companies are currently in a state where they cannot venture producing new products and only then search out the buyer for them. It is objectively necessary to work to order, for the concrete consumer. That is why, after the technical task is formulated, intellectual potential is focused on the resolution of concrete technical questions. If an idea for a new technical decision arises on the way, it is offered to the customer as a new variant, and frequently this is the way the innovation is realised. Founders of an RSO company conduct R&D activity themselves. Workers of partner manufacturing enterprises are brought in to execute necessary work on the level of detailed manufacture and assembly of the equipment (often for reducing the costs of the order which are to be paid in money terms). The aim of further R&D is to work out the most economic (from the viewpoint of consumables, energy and time) regimes of processing, and provide the reliable characteristics of final products (such as depth and firmness of hardening).

Work for the concrete customer with his sophisticated needs means the share of additional R&D work being invested in the outcome product in practically each case. Moreover, the period of accomplishing the order is much shorter – for various technologies it constitutes from 3-4 to 60 to 90 days from achieving the technical task to its execution. Earlier, when researchers have worked as employees of the institute, the whole path has been prolonged up to 1.5-2 years because of paperwork and the large number of authorities it had to get through.

The second, external factor influencing the time lag between the appearance of an innovative idea and issue of a new product on the market is demand for a particular product. One situation is a bright illustration of the stimulating role of the market need for the introduction of new scientific and technological developments. Not being practically claimed within many years, the idea was realised and prepared for industrial introduction at the level of production of 300,000 bearings from innovative material per one year, when a demand became apparent, within two weeks. This concrete innovation became urgently necessary for one of the major factories of Belarus, as deliveries of raw material from Russia previously used at this factory for producing bearings – bronze – had stopped after separation of both countries. Because an urgent necessity arose to find the substituting material (otherwise the factory would be closed), an expert report in Moscow was carried out in *two weeks* instead of the usual 2-3 years. The patent was filed, and manufacture of such necessary innovative material was prepared on a large industrial scale.

Linkages of scientists of high educational establishments and scientific research institutions with industry have previously been formed under the framework of executing the particular tasks in *scientific and technical programmes, separate projects, state budget and economic contracts*. In 2002, 842 scientific developments of universities were exploited in industry – 40 new kinds of machines, equipment and devices, 33 kinds of materials and substances, 81 technological process, 16 systems, set-ups, 3 kinds of plants and drugs have been developed and introduced.

Still, 75% of enterprises do not collaborate with scientific and research institutions at all (Metolit 2003). Those which do, do collaborate first and foremost with public research institutes (9), NASB (6), and only 4 with universities. This preference is based chiefly on the high reputation of the public research sphere in Belarus, which can be used as a good ground for the promotion of research-based spin-offs which originate from these institutes. As qualitative analysis (Gordienko, 2001) shows, it really is the case. Moreover, having built contacts with industrial partners on the basis of the trademark of 'parents', spin-off firms 'pay the debt' by their own developments (which are quicker, oriented to very specific demand, and are relatively cheap), and industrial enterprises are more eager to cooperate with public research institutions in joint research projects with the longer period of realisation – they are confident of the effectiveness of such collaboration.

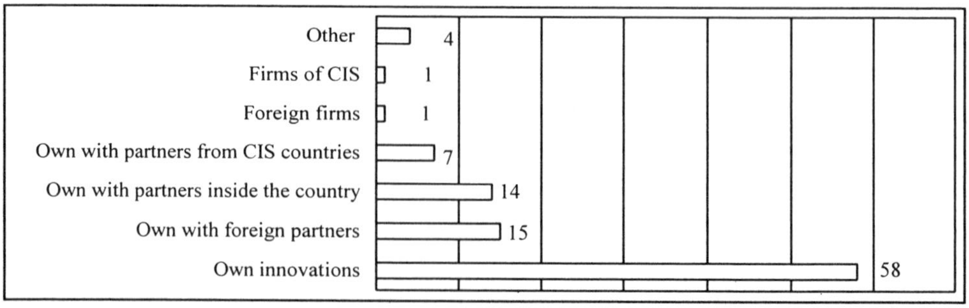

Fig. 1: Technological cooperation: distribution of schemes, 2002
Source: Metolit, 2003

Though intensive attempts to cooperate with "government" (SSTP) have been made, they all have proved to be in vain. Taking part in state programmes entails big problems, especially of a financial and bureaucratic character. The experience of participating in numerous state programmes is accumulated, but all have finished with identical results. The way the programme was financed caused complete failure of the programme performance. One is not allowed to involve private (individual) customers in the state project, who would actually be prepared to finance the developments. This results in a breach between the spheres of the academic (institute) science, industry, and the private (individual) initiative. Technological collaboration of RSOs with *parent research institutions* is important for the latter because it is too difficult for the large research institute where overhead expenditure constitutes about 130-150% to materialise the innovation, and the products thus become too expensive.

Innovators constantly cooperate with former collaborators from the parent research institute; scientific conferences help seriously in communication and studying the new prospects. Contacts with all developers in the region are established. Often enterprises are members of scientific societies in their technological area. Because RSOs are tackling the technological issues that belong to several overlapping technological fields, experts from other enterprises and research organisations are regularly involved in the execution of concrete jobs (for example, experts on thermal processing, experts on design, research, technologies and equipment development). Besides, collaborating structures are organised, as temporary scientific groups (VNK), where the brainstorming potential from various organisations, also research institutes, comes together with a view to finding "cross-cultural" solutions.

Overlapping technologies lead to tight *integration of the elite researchers' community*. There are also cases where innovators help one another in technological collaboration informally, when they, for example, let the other RSOs make tests on their (unique) installations, and even make tests by themselves. Sometimes they even pass the installations they have made for their innovative technological process to "enterprises/plants which are manufacturing the same products as we are" (both in Belarus and Russia) – as the researcher-innovators call them. "A mere apology for competitors they are – we collaborate; they are not able to compete, they have no experts, they are practically bankrupt, and equipment we have developed is passed to them on a collaborating basis, to keep them afloat". If there exist firms with a similar profile of activity (also in Russia or Ukraine), cooperative relations are established with them.

Technological cooperation *with foreign research organisations* is still not successful, because of the conflict of interests. Usually they possess larger funding opportunities for conducting collaborative R&D projects, whereas the Belarusian party contributes intellectual capital, but the foreign party expects the larger share of intellectual property on the basis of their investments. Cooperation with *foreign business* is subject to the same conditions, but there have already been examples of successful cooperation with them, based on the organisation of joint production with an exclusive right to sales on the market of the foreign partner. Conditions are not always fair, but Belarusian entrepreneurs have to accept them because the possibilities for financing the development of their activity are highly restricted.

V. Conclusions. Vision of the RSO future

Processes of privatisation in the scientific sphere of Belarus take place very slowly, and the quantity of firms, which objectively are able to bring to the economy the innovative technologies, is very restricted. These enterprises cannot be numerous in the economy, and understanding of this fact on the political level is important. But even more important is the high value that each of them contributes to the innovation system. Incremental growth in one unit of these firms gives the rough push to the whole cluster of related technologies, many of them leading to innovative products, and some of them belonging to the high-tech sphere by themselves.

"Such enterprises as ours, bringing innovative technologies to our industry, are the hens which lay gold eggs for the state – and the authorities want to cook the hens" – one respondent says. If nothing changes in current taxation, general economic situation, and attitude to small innovative entrepreneurship, the only option to enable Belarusian research-based spin-off companies to survive is to transform them into companies with a share of foreign capital, because joint and foreign enterprises are more favourably treated by legislation than domestic small innovative enterprises.

Emergence of a stratum of innovating-for-the-market researchers appears to be a valuable linking chain strengthening the weak linkages "university-science-industry". However, their role is not merely restricted to this commonly accepted function of technology transfer. The innovation competencies and culture fostered by them in an innovation system serve as the catalyst for developing a knowledge-based economy.

References

Development of Science and Technological Potential in Ukraine and Abroad. (1996). Collected volume of analytical materials. Issue 2(10). Centre for Research of Scientific and Technological Potential and History of Science (CIPIN) after G.M. Dobrov, National Academy of Sciences of Ukraine, Kiev.

Development of Science in Belarus in 2002. (2003). (Analytical report). Ivanov, V.F., Artiuhin, M.I., Grigianets, R.B., Jenin, S.V., Karelina, V.A., Korshunov, A.I., Kostiukovitch, N.I., Meerovskaja, E.D., Slonimski, A.A., Hartonik, I.A. Committee for Science and Technology. Minsk, BelISA.

Eisenhardt, K. (1989). Building Theory From Case Studies. In Academy of Management Review, no. 14 (4), 532-550.

Financial and economic provision of innovation activity (2003) Materials of seminar on 18 December 2003. Technopark BNTU "Metolit".

Gordienko, A. I., Pobol, A.I., Pobol, I.L. (2002). "Spin-off Firms Around the Academic Institute as a Form of Innovation Networks Development in Belarus". In: Actual Issues of Innovation Activity Development. VII International Scientific and Practical Conf., Alushta, Ukraine, 16-21 Sept. 2002, P.185-192.

The Inter-University Centre for Marketing of Scientific-Research Works (ICMSRW) – http://www.icm.by/

Law of Republic of Belarus "About the state support of small entrepreneurship in Republic of Belarus" of 16.10.1996 № 685-XIII.

Lundvall, B.-Å., Johnson, B., Andersen, S.E., Dalum, B. (2002). National Systems of Production, Innovation and Competence Building. In: Research Policy. Vol. 31, Elsevier Science.

The National Centre of Legal Information of the Republic of Belarus – http://www.ncpi.gov.by/, http://president.gov.by/Minstat/en/indicators/industry.htm/ – Main statistics of economic development of Belarus

Nelson, R.R. and Nelson, K. (2002). Technology, Institutions, and Innovation Systems. In: Research Policy. Vol. 31, Elsevier Science.

Niosi, J. (2002). National Systems of Innovation Are x-Efficient (And x-effective). In: Research Policy. Vol. 31, Elsevier Science.

Chapter 8
Innovation at the Local Level – Trends in Bulgaria

PLAMEN SVETOSLAVOV GRAMATIKOV
Southwest University "Neofit Rilsky"
66, Ivan Mihailov Blvd.
BG-2700 Blagoevgrad
Bulgaria

Summary

This chapter discusses some innovation tendencies in the European countries and in the USA. Special attention is spent on the Bulgarian trends, especially at the local level.

Introduction

The relationship that manufacturing firms build with other organisations such as customers, manufacturing suppliers, producer service providers, competitors and research institutions enable them to deal with changes in technologies, markets and other aspects of the business environment.

Technology transfer is the movement of technological and technology-related organisational know-how among partners (individuals, institutions and enterprises) in order to enhance at least one partner's knowledge and expertise and strengthen each partner's competitive position. Technology transfer occurs throughout all stages of the innovation process. It can take place via informal interactions between individuals, formal consultancies, publications, workshops, personnel exchanges and joint projects, involving groups of experts from different organisations, and more readily measured activities such as patenting, copyright licensing and contract research.

The network approach of innovation research departs from the single-act philosophy of innovation and views technological innovation as a social process as well as a technical process. This innovation process involves interaction between individuals within the system, both internally within the firm and between members of the firm and outside organisations. Technological innovations are increasingly created, developed, brought to the market and diffused through complex mechanisms built on linkages between different organisations. Some authors (Hakansson, 1987) attempt to explain this process as network relationships between these various actors. The process could be illustrated by Figure 1.

The absorption capacity of a firm, i.e. the ability to learn, assimilate and use knowledge developed elsewhere through a process that involves substantial investments, crucially depends on the learning experience which in turn may be enhanced by in-house R&D activities. Thus, R&D may be seen as serving a dual but strongly interrelated role: first, in developing new products and production processes and secondly, in enhancing the capacity to learn.

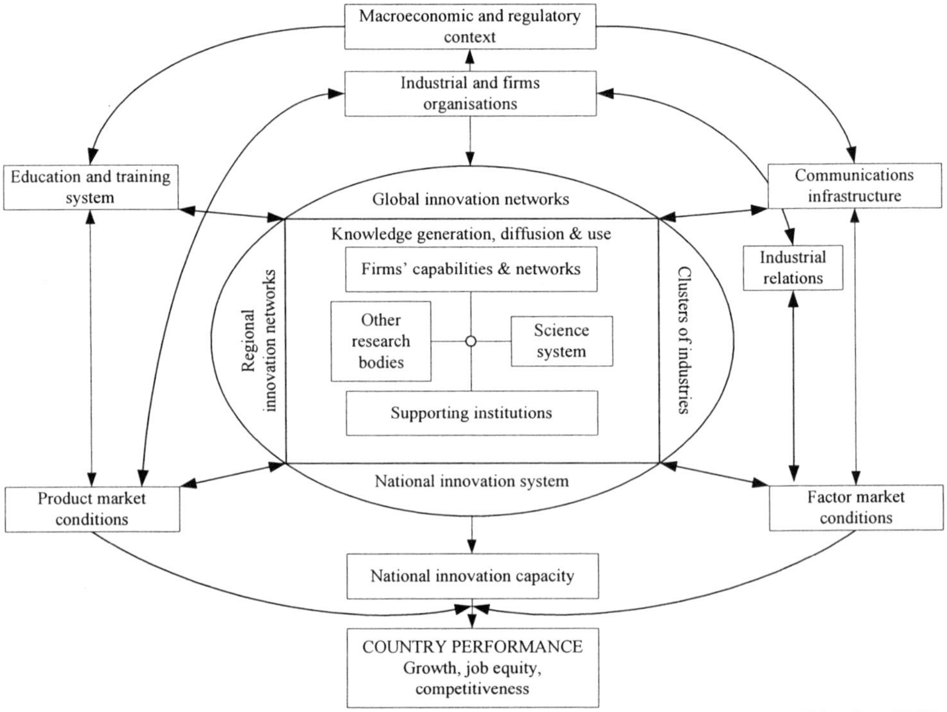

Taken from OECD

Fig. 1: Basic actors and links in an innovation system

Innovation networks

Innovation ideas can arise by reason of many and different sources in each investigation stage as development of a product, its manufacture and marketing. Availability of a policy and stable research, technological and developing infrastructure in the science area and in the technologies do not lead automatically either to creation of innovation business, or to the automatic increase of the competitiveness of the economy, as seen in Figure 2:

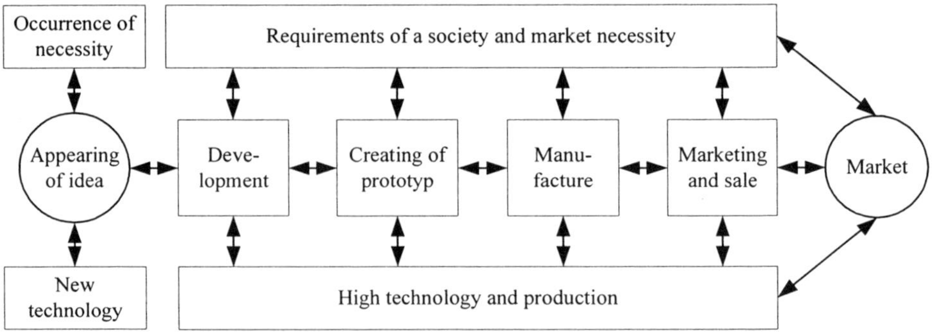

Fig. 2: Model of an Innovation Business

Firms, especially smaller firms, that lack appropriate in-house R&D capacities have to develop and enhance their absorption capacity by means of other sources such as by learning from customers and from suppliers, by interacting with other firms through the creation of innovation networks and by taking advantage of knowledge spillovers from other firms and industries. These sources provide the know-why, know-how, know-who, know-when and know-what important for entrepreneurial success (Malecki 1997).

There are different definitions of *innovation networks* (DeBresson and Amesse 1991, Freeman 1991). One of them defines a network as "an evolving mutual dependency system based on resource relationships in which their systematic character is the outcome of interactions, processes, procedures and institutionalisation. Activities within such a network involve the creation, combination, exchange, transformation, absorption and exploitation of resources within a wide range of formal and informal relationships" (Tijssen 1998).

Interfirm collaboration is an important source of knowledge, which can be organised in five types of network (Fischer, Suarez-Villa and Steiner 1999):

1. *Customer networks* which are defined as the forward linkages of manufacturing firms with distributors, marketing channels, value-added resellers and end-users.
2. *Manufacturing supplier networks* which are defined to include subcontracting, arrangements between a client and its manufacturing suppliers of intermediate production inputs.
3. *Producer service provider networks* that are defined to include arrangements between a client and its producer service partners (esp. computer and related service firms).
4. *Producer networks* which are defined to include all co-production arrangements that enable competing producers to pool their production capacities, financial and human resources.
5. *Cooperative relationships with research institutions or universities* (pre-competitive stage) pursued to gain rapid access to new scientific and technological knowledge and to benefit from economies of scale in joint R&D.

It is interesting also to consider the role of geography in innovation. Why does innovation cluster in space? It can be seen in Figure 3 that spatial proximity enhances the efficiency of innovation-based interactions:

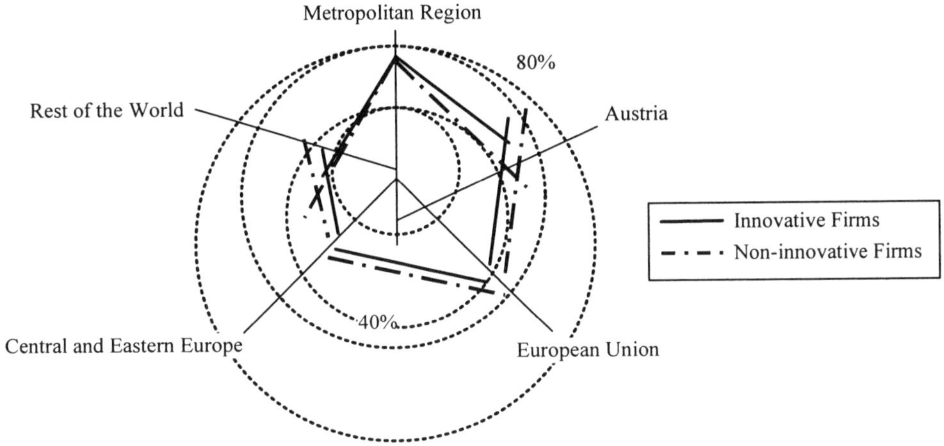

Fig. 3: The Geography of Innovation Networking (percentage of networking firms of all firms)
Source: Vienna innovation survey, 1998

Innovation systems in the USA and in Europe

a) USA (Source: Department of Commerce, Tech Transfer 2000: Making Partnerships Work)

Federal Government Technology Transfer Support:

- Share information
- Commercialise Federal lab R&D
- Foster Federal lab – business interactions
- Foster linkages across sectors: Business – Business; Business – Universities; Business – University – Federal labs

Legislation:

- Improve the environment for collaboration
- Commercialise public-funded R&D
- Support for small and medium-sized enterprises (SMEs)

Mechanisms for Federal Technology Transfer:

- Collegial interchange, workshops, conferences
- Consulting
- Work for others and use of facilities
- Exchange programs
- Technical assistance
- Information dissemination: National Technology Transfer Centre, Regional tech transfer centres, Federal research in progress database, Start-up or spin-off companies, Cooperative research and development agreements (CRADA), Licensing.

Input/Investment Indicators:

- R&D expenditure by sector
- Industrial R&D expenditure
- Government R&D funding
- Academic research and R&D infrastructure
- Science and engineering degrees
- Science and engineering workforce

Output/Process Indicators:

- Publication and citation patterns
- Patenting
- Innovation
- Trade
- Public attitudes & understanding

The US Budget System and Its Political Context:

- Budget-main tool for setting priorities, planning, managing
- R&D part of regular budget process
- Balance of power between President and Congress, neither has full control
- System is fragmented, often chaotic, unpredictable

US Organisation for Science and Technology:

- No unified R&D budget
- No Department of Science and Technology
- More than 25 departments and agencies conduct R&D
- Most R&D supports agency mission
- Some departments have several units, each with own R&D program

American Grant and Contract System:

- Most government-funded R&D conducted *outside of government* under grants and contracts
- Most funds are competitively awarded
- Basic research is done in universities; grants use merit review
- Applied research and development mainly in industry under contracts

U.S. Priority-Setting:

Macro-level

- R&D priorities derive from national priorities
- Politics, not science, is the key
- Policy-makers concerned with agencies, large programme (not individual projects). Exception – earmarked appropriations (e.g. pork barrel)

"Meso"-level

- Agency (e.g. NSF, NIH) budgets are appropriated by major units/programs
- Mix of political and scientific factors
- Administrative responsibility delegated to lower levels
- Higher-level administrators come in if problems arise

Micro-level

- Choices made among competing projects
- Merit (peer) review, scientific factors most important at this level
- Priority-setting takes place within framework of higher-level priorities
- Politics at this level can damage R&D

b) *Most European countries* (Source: Albert H. Teich, AAAS; CIPRE – Budapest, 2002)

The Budget System and Its Political Context:

- Most have Parliamentary systems
- Budgets are done within the government
- Majority means virtually automatic approval
- Less opportunity for influencing outcomes

S&T Organisation:

- Many have department or ministry of science, often combined with education (or other functions)
- R&D often more centralised
- One or two agencies support basic research

R&D Systems outside the US:

- In general, a larger share of government R&D is done inside government

- Competitive review less widely used
- More basic research done in research institutes
- Applied research in governmental agencies

National and Regional Innovation Systems

National Innovation System (NIS) is "a set of institutions whose interactions determine the innovative performance of … national firms" (Nelson 1993).

The technological distinctions between the technically advanced countries and other states are result not only of technological innovations but also of a level of organisations, financing, marketing, and education and of other factors. The combination of all factors that contribute to overcome these differences is, in fact, a national system of innovation.

These links between all participants are shown in a simplified model of NIS in Figure 4. In this figure research and educational institutes are connected with firms to transfer skills and experience. The intermediaries and the elements of educational infrastructure play an essential role in this transfer. The market demand is a basic motor for the firms.

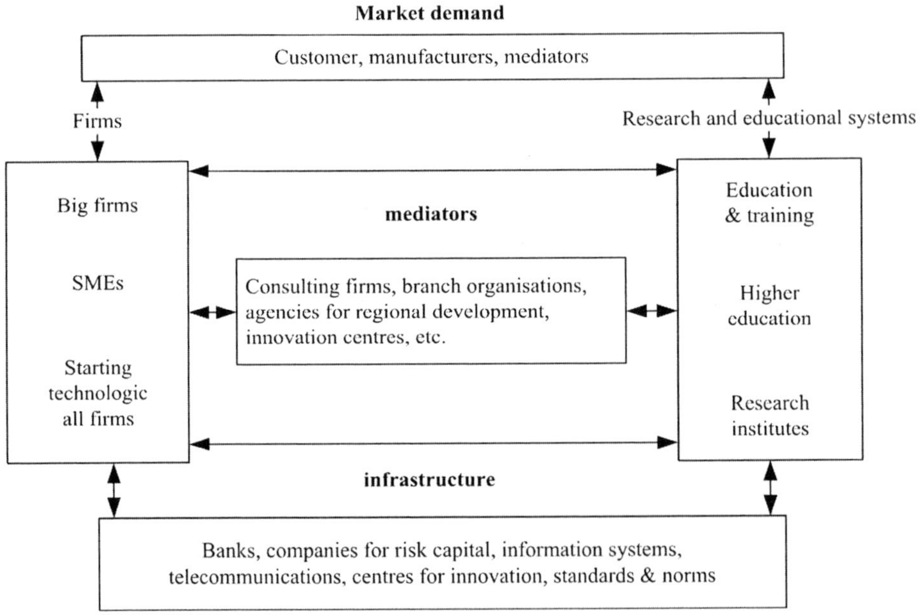

Fig. 4: Simplified Model of National Innovation System

A regional dimension of an NIS is a *Regional System of Innovation* (RIS). While the NIS defines rules, institutions (taxation, education system), research financing, legislation etc., RISs (Figure 5) carry out interactions between firms and universities at local level.

Beginning in the 1980s a new set of interventions emerged to encourage the development of regional innovation systems both in the USA and in the Europe. The first successful RISs such as Silicon Valley in California, Route 128 near Boston (USA) and Cambridge (UK) have developed autonomously and have induced high economic growth. It can be concluded that there is no "ideal model" of regional innovation policy (Gramatikov, 2003).

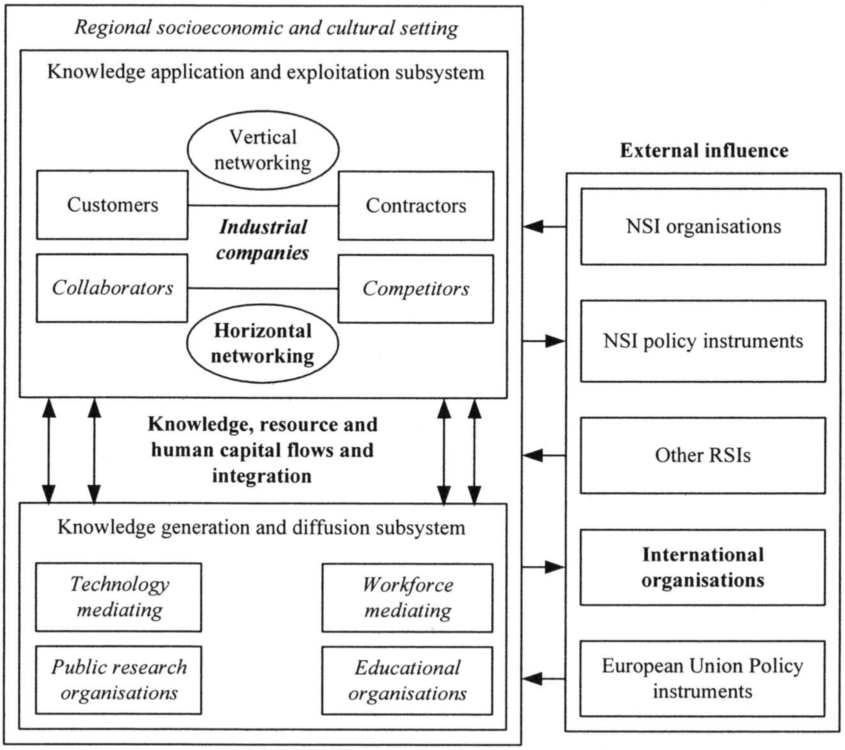

Fig. 5: A Model of a Regional Innovation System
Source: Attila Varga, University of Pecs; CIPRE – Budapest, 2002

Regional innovation policy supports:

- Creation of new knowledge
- Exploitation of new knowledge
- Absorption capacity
- Diffusion of new knowledge
- Awareness building among less innovative companies

1. Regional innovation policy in the USA:

- Main policy making units are US state governments
- Typology of programmes:

a) Technology development (research & applications)
- university–industry technology centres
- government–industry consortia
- university-industry research partnerships
- equipment and facility access programmes

b) Industrial problem solving
- technology extension
- federal technology application programmes

- implementation grants

c) Technology financing
- company financing
- project financing

d) Start-up assistance
- Incubators
- research parks

e) Promoting industrial network building

2. Regional innovation policy in the EU:
- Performed both at the regional and/or national level
  - provision of R&D subsidies
  - university funding
  - support for technology centres
  - encouraging research-industry networking
  - innovation support for SMEs
  - public support for risk capital and innovation financing
  - education and training
- Performed by the European Commission
  - European Structural Funds to support economic development
  - European Framework Programmes to support regional technology development
  - Regional Technology Plants (RTPs)
  - Regional Innovation & Technology Transfer Strategy projects (RITTS)
  - Regional Innovation Strategy (RIS) projects

Innovation Trends in Bulgaria

Every country and every government realises its R&D policy, the written or unwritten one, every year during the preparations and approvals of the state budget. European countries of a medium size without significant mineral resources must base their prosperity and the satisfaction of their citizens first of all on a fast and efficient gaining of new knowledge and its utilisation in new products, technologies and services.

The Republic of Bulgaria, as a medium sized European country trying to be accepted into the EU, must understand these trends. Unfortunately the starting point for financing R&D in Bulgaria is not so good. According to the National Statistical Institute only about 0.52% of the Bulgarian GDP was spent on R&D during 2000. The business sector's share for R&D totalled 0.11% of GDP while 0.41% was financed by the government budget.

According to Eurostat Bulgaria is 9th in range among the 11 countries that are candidates for EU membership. The financial expenditure on R&D in these Central and East European countries is shown in Table 1. It can be seen that Bulgaria occupies one of the last places, before Latvia and Romania, in this classification.

R&D participation in meeting the prospective needs of people and the society, including the economy, requires a wider spread of the R&D capacities' division. This is specially felt in regions, in which a wide restructuring of the industry takes place, and will continue to take place. The provisions for the support of the transfer of R&D results will be proposed in such a way that they will influence the situation in the regions efficiently.

Table 1: Expenditure on R&D

	Total expenditure for R&D, % of GDP	**R&D Expenditure of the National Business, % of GDP**	**Year**
Bulgaria	0.52	0.11	2000
Czech Republic	1.35	0.81	2000
Estonia	0.68	0.15	2000
Hungary	0.80	0.36	2000
Lithuania	0.60	0.07	2000
Latvia	0.48	0.20	2000
Poland	0.70	0.25	2000
Romania	0.41	0.30	1999
Slovenia	1.51	0.83	1999
Slovak Republic	0.69	0.45	2000
Turkey	0.63	0.24	1999

Source: Eurostat

It is interesting to compare some basic indicators for enterprise groups in the EU and Bulgaria (Table 2):

Table 2: Comparative indicators for enterprise groups in the EU and Bulgaria

Indicator		**SMEs**				**Large enterprises**	**Total**
		Micro	**Small**	**Medium**	**Total**		
Number of enterprises ('000)	EU	18 040	1 130	160	19 330	38	19 370
	Bulgaria	190	11	4	205	0.8	206
Number of employees ('000)	EU	38 360	21 320	14 870	74 550	38 680	113 230
	Bulgaria	342	242	391	975	705	1 680
Average number of employees in one enterprise	EU	2	20	90	4	1 010	6
	Bulgaria	2	22	98	5	881	8
Average sales per enterprise (EURO '000)	EU	0.2	3	23	0.5	215	1.0
	Bulgaria	0.02	0.33	1.83	0.06	11.37	0.1
Export share in sales (%)	EU	6	13	16	11	22	16
	Bulgaria	7	7	13	9	22	15
Value added per employee (EURO '000)	EU	30	50	95	45	90	60
	Bulgaria	1.1	1.8	4.9	2.0	1.8	1.9
Labour costs share in value added (%)	EU	40	53	43	45	38	42
	Bulgaria	33	64	72	56	64	62

Source: The European Observatory for SMEs, Sixth Report. European Commission, 2000. NSI, 2000.

Bulgaria requires policies in the field of science, technology and innovation with the following *purposes*:

1. Increase of competitiveness of the Bulgarian industry through the introduction of scientific attainments, technology and innovation.
2. Consolidation of scientific and technological sector through their cooperation, concentration and strengthening of links with industrial sector.
3. Creation of favourable conditions for young experts in the field of science and technology.

To achieve these purposes it is necessary to pass *4 steps:*

1. Development of a national innovation system, which will cover all institutions, firms and organisations working in the field of development, exchange and application know-how.
2. Maintenance of cooperation between scientific and technical sector and business. This cooperation is connected with the interaction between the ministries of economy, education and science, etc.
3. Creation of a national organisation with the purpose of coordinating acceptance and performance of a strategy of innovation. This includes a National Innovation Council with a National Executive Body.
4. Provide for the adequate financing of innovation.

Bulgarian National Innovation Strategy suggests 11 concrete steps and 3 activities:

Steps	
1	Optimisation of the system "science – technology – innovation"
2	Creation of a grant system for the financing of fundamental scientific research
3	Creation of a partial grants system for new technologies and products
4	Strategic financing of adjunct professors and senior researchers
5	Stimulating cooperation between public sector and private sector by creation of centres of competence
6	Creation of Advisory Councils at the universities and faculties
7	Encouragement of the employment of young specialists
8	Creation of units for innovation transfer in agriculture and food industry
9	Establishment of an Association of Innovation Firms
10	Stimulating foreign investments in R&D
11	Creation of technological parks
Activities	
1.	Campaign for public knowledge
2.	Creation of a National Innovation Council
3.	Creation of a National Executive Body for innovation

The Bulgarian National Priorities and R&D Programs are:

- Information Society – communications and computing, digital content and intelligent interfaces, e-services for citizens and business
- Genomics – plant genomics, bioinformatics, proteonics
- Nanotechnologies and New Materials – nanomaterials, ultrathin layers, multilayer nanosystems, molecular design
- Bulgarian Society – Part of Europe and the World (Effective participation in the European administration area, catch-up of differences between regional economic development)

- Cosmic Research – remote control and elaboration, study on cosmic weather
- Science and Society, Sustainable Development, Global Change and Ecosystem

The Structure of the Research Sector in Bulgaria:

- Universities – 37 universities with about 9,100 researchers/lecturers
- Bulgarian Academy of Sciences (BAS) – encompasses 67 research units (institutes and labs), about 3,600 researchers
- National Centre for Agricultural Sciences (NCAS) – 26 research units, about 2,900 researchers
- Others (medical, industrial, private institutes), approximately 1,100 researchers

The statistics about enterprise density by district (Table 3) show the increased importance of big cities for the growth in the number of companies in the respective district.

Table 3: Enterprise density by district

DISTRICT	1998		1999	
	per 1000 people	**per 1000 people of the active population**	**per 1000 people**	**per 1000 people of the active population**
Total for Bulgaria	**24.09**	**43.27**	**25.80**	**44.46**
BLAGOEVGRAD	26.20	44.50	26.88	45.25
BURGAS	27.20	46.78	34.32	58.53
VARNA	29.70	50.29	33.87	56.68
VELIKO TARNOVO	25.00	44.29	23.66	41.35
VIDIN	19.70	38.85	20.31	39.23
VRATSA	16.50	30.61	12.63	23.15
GABROVO	31.40	55.58	30.81	54.26
KARDJALI	16.00	27.76	16.73	27.63
KYUSTENDIL	26.00	46.75	26.80	47.87
LOVECH	26.50	49.95	26.11	48.74
MONTANA	18.60	36.26	18.91	36.34
PAZARDJIK	21.30	37.19	21.26	36.86
PERNIK	18.40	32.84	19.98	35.39
PLEVEN	25.20	46.77	26.03	47.77
PLOVDIV	26.80	45.55	23.80	40.71
RAZGRAD	18.10	31.20	16.75	28.72
ROUSSE	26.90	46.65	30.54	52.60
SILISTRA	21.10	36.55	19.29	33.13
SLIVEN	24.40	42.34	24.92	43.42
SMOLYAN	30.30	51.47	32.16	52.60
SOFIA – CITY	26.90	42.71	31.47	50.47
SOFIA – DISTRICT	20.60	38.36	20.11	36.66
STARA ZAGORA	27.90	48.66	24.02	41.83
DOBRICH	23.30	39.52	24.72	41.88
TARGOVISHTE	21.50	38.19	19.61	34.75
HASKOVO	29.50	51.49	28.36	50.75
SHUMEN	23.90	41.15	22.78	38.96
YAMBOL	25.50	46.31	25.01	45.07

Source: NSI and in-house calculations.

A key instrument for the application of Bulgarian state policy for regional development in the context of the Regional Development Act is the *National Plan for Regional Development* (2000-2006). It defines three priorities further elaborated in the district development plans: creation of conditions for sustainable development of individual regions, reducing regional differences in employment, incomes and growth potential and opening of the national economy through development of cross-border cooperation.

A substantial part of R&D capacity is centred in Bulgaria in several larger cities and a few large traditional universities. The balance of the economic structure as a whole, the absence of imbalance in industrial development, the export-orientation of SMEs, a greater adaptability to market development and the implementation of innovations positively influence the conditions for SME development in the regions. Large cities are more attractive with respect to investment, including direct foreign investment.

Burgas, Varna, Sofia city and Rousse are the districts that registered the greatest increase in company density for one year. Stara Zagora and Plovdiv, however, are the districts showing a decrease in company density, which indicates that the existence of a big city as a district centre is not enough. It is obvious that the factors for business start-ups are much more complex.

Comparisons of the list of the ten districts with the greatest and the smallest density of enterprises in 1999 and 1998 show no significant changes. Stara Zagora and Plovdiv lost their good positions, while the other big cities moved from the second to the first half of the list of the ten districts with the highest enterprise density (ASME 2001).

One of the most important factors for innovation is the technological equipment of the firms. The average age of equipment in Bulgaria can be seen on Figure 6.

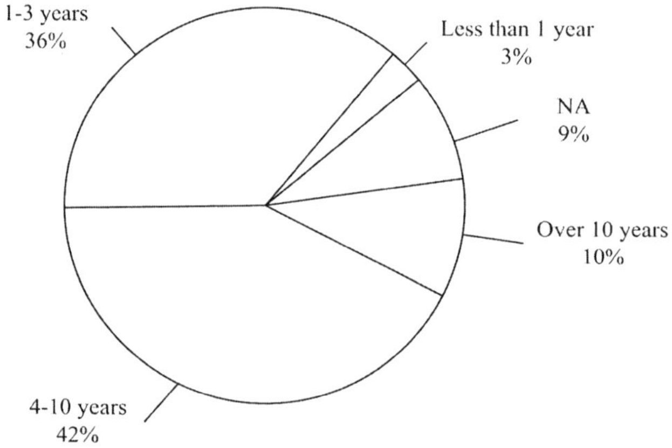

Fig. 6: Average age of equipment

Source: Project "Dynamic entrepreneurs: comparative study of Bulgaria and Poland", Centre for Entrepreneurship Development, University of National and World Economy, Sofia, 1999-2000

There were significant differences by region with respect to the amount of investment and especially the resources dedicated to the purchase of new equipment. In 46.2% of the cases the SMEs interviewed reported that the average age of their basic equipment was within the range of 4–10 years; 23.7% reported an average age of equipment of over 10 years and 2.9% of the enterprises possessed equipment which was less than a year old (See Figure 7).

In spite of this, SMEs in the regions in industrial decline most frequently reported that they had equipment less than one year old – in 6.9% of the cases. This fact was due to the highest percentage (18.8%) of companies established during the past two years (1999-2000) in these regions. 45.2% of all companies started in the last two years were located in the regions in industrial decline.

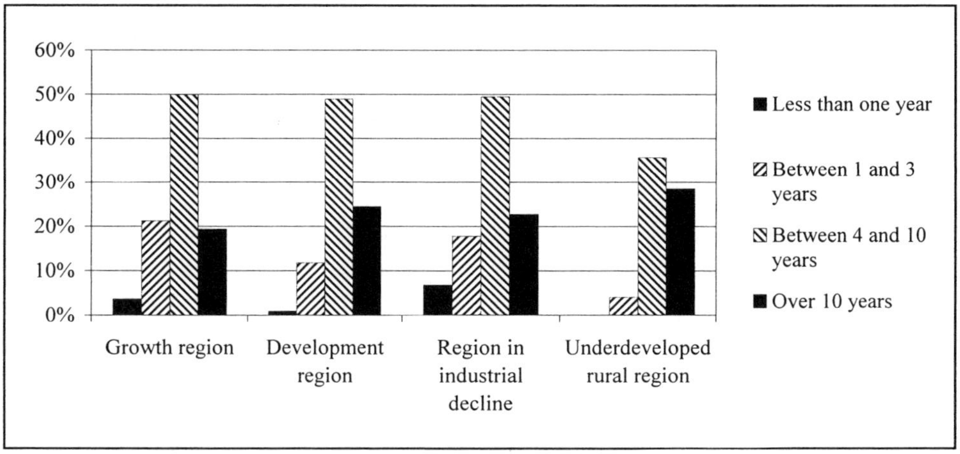

Fig. 7: SME equipment age by region of purposeful intervention
Source: sample survey

The barriers to the innovation of new, competitive products are, first, the lack of financing – 61.8%. Next is inappropriate production equipment – 23.7%, and the lack of knowledge about new technology – 21.1%. (Figure 8).

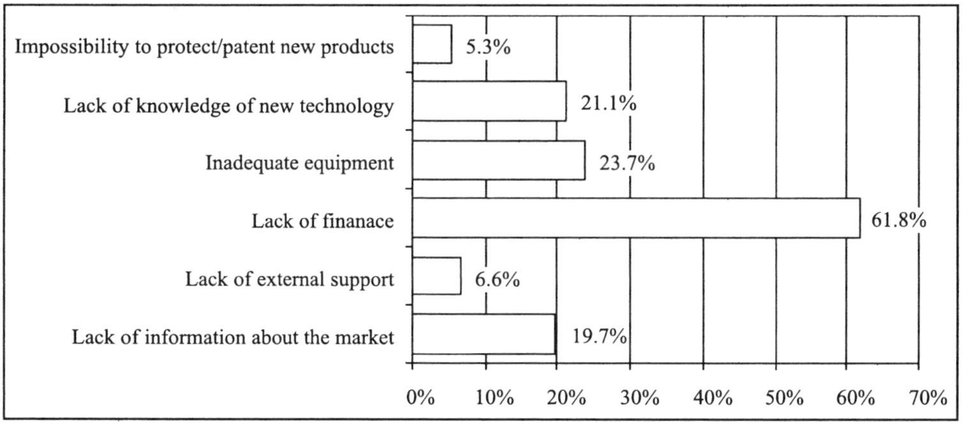

Fig. 8: Main barriers to new product development
Source: Project "Dynamic entrepreneurs: comparative study of Bulgaria and Poland", Centre for Entrepreneurship Development, University of National and World Economy, Sofia, 1999-2000

Conclusions

Regional differences to a certain extent reflect the problems of SMEs in Bulgaria. Regional problems, including those related to SME development, can be addressed only through complex measures. Capacity building for SME development is the most effective and relatively easy to implement regional policy.

SMEs in the different types of regions registered a substantial growth in sales revenue, except for the SMEs in the regions in industrial decline. Generally, the SMEs in all regions relied extensively on their local markets (the municipality or town). More than two thirds of the enterprises in all regions had formulated plans for growth in the short term (for the next year).

Attracting customers and ensuring start-up capital were important issues for companies at the start-up stage. Only companies in the underdeveloped rural regions and the growth regions reported the lack of skilled workers as a problem.

The type of information that the Bulgarian SMEs sought most often was on domestic markets, tax information, information about suppliers and accounting information.

The entrepreneurs in the different regions are not only working in different conditions, but they also have different value systems and attitudes, or, they have a different entrepreneurial culture. In this sense, future programmes and measures for SME promotion should if possible take into account these differences with a view to increasing SME efficiency.

References

DeBresson C. and Amesse F. (1991) Network of innovators: A review and introduction to the issue, Research Policy, 20, 363-379.

Fischer M. M., Suarez-Villa L. and Steiner M (Eds) (1999) Innovation, Networks and Localities, Springer, Berlin, 11-27.

Freeman C. (1991) Network of innovators: A synthesis of research issues, Research Policy, 20, 499-514.

Gramatikov P. S. (2003) Regional Innovation Systems – Key to the Technological Development, University Review Journal, South-Western University Press, Blagoevgrad, 1, 11-17.

Hakansson H. (1987) Industrial Technological Development: A Network Approach. Croom Helm, London.

Malecki E. J. (1997) Technology & Economic Development, Longman, Essex (2nd edition).

Nelson R. R. (ed.) (1993) National Innovation Systems. A Comparative Analysis, Oxford University Press, New York and Oxford.

SME Report 2000; ASME, Sofia, 2001.

Tijssen R. J. W. (1998). Quantitative assessment of large heterogeneous R&D networks: The case of process engineering in the Netherlands, Research Policy, 26, 791-809.

R&D Priorities in Innovation Policy and Financing in Former Socialist Countries
W. Leal Filho and P.S. Gramatikov (Eds.)
IOS Press, 2005

Chapter 9
The Determinants of Poland's Innovative Capacity

MARZENNA ANNA WERESA
Warsaw School of Economics
Al. Niepodleglosci 162
02-554 Warsaw
Poland

Summary

This paper uses the concept of national innovative capacity in order to assess Poland's performance in the creation and diffusion of new knowledge. Evidence from aggregate patent and R&D statistics suggests that Poland's innovation performance has been relatively weak compared to other transition countries from Central Europe. A scarcity of financing is the most limiting factor for R&D development in Poland. In the period of 1989-2003 R&D as a share of GDP amounted to less than 1%. The primary R&D efforts of Poland remain concentrated in state institutions. Governmental funds finance the majority of R&D expenditures in Poland. Without an increase in private R&D financing, it will be hard to close the technology gap.

Moreover, this paper identifies the pattern of Poland's technological specialization using the RTCA index. It shows that Poland has comparative technological advantage in the medium-high sector, in particular in the motor vehicle industry and chemical industry, and in medium-low technology industries, such as metallurgy and manufacturing of some sport equipment and simple machine parts. The pattern of Poland's technological specialization has been changing since the beginning of the transition process.

Introduction

The rapid changes in science and technology that are observed nowadays mean that economies have become more and more knowledge-based. The ability to innovate, or at least to get access to new knowledge and imitate innovation, seems to be increasingly important. Technology and innovation have become a crucial factor in economic growth. Therefore, there is a great deal of debate concerning the underlying drivers of the innovation process[1]. The latest theoretical approach developed by S. Stern, M. Porter and J. Furman introduces the concept of *national innovative capacity*, which is defined as the ability of a country to produce and commercialize a flow of innovative technology over the long term (Stern, Porter & Furman, 2000, p. 1). This framework is an attempt to integrate macro- and microeconomic perspectives regarding the sources of innovation at national level. It allows examining Poland's performance

---

1 Innovation is defined here in a broad Schumpeterin sense, as changes in engineering design, capabilities of production teams and their methods of work (including organizational and marketing innovations). The recent literature points out the growing convergence in the use of the concepts of innovation and technological change, treating them as synonymous (Cantwell, 1996; Chen, 1996).

in science and technology and comparing it with the EU average and with other transition economies. These findings have also some policy implications. They indicate areas which should be developed in technological policy agenda in order to increase Poland's national innovative capacity in the future.

The concept of national innovative capacity

What makes firms innovate? Why do some countries innovate more that others? What role do national expenditures on R&D play in creating innovative capabilities? Does business and institutional environment matter for innovation performance? In order to answer these questions the concept of the national innovative capacity has been developed (Stern, Porter & Furman, 2000). It identifies the determinants of the innovation process by drawing on three distinct strands of prior research:

- models of ideas-driven growth (Romer, 1989);
- Michael Porter's concept of industrial competitive advantage based on the industrial cluster approach (Porter, 1990);
- the innovation systems approach (Nelson & Rosenberg, 1993).

All three perspectives yield common insights into the creation and commercialization of new knowledge. Their integration indicates that the determinants of national innovative capacity can be divided into three broad areas (Stern, Porter & Furman, 2000). The first is the strength a nation's common innovation infrastructure, which includes R&D expenditure, investment in higher education and funding of basic research, the number and quality of scientists and engineers as well as a country's science and technology policy. The second area is defined as 'cluster-specific innovation environment'. It is proxied by the extent of financing of R&D by a country's private sector. And, finally, the third set of determinants concerns the quality of linkages between the common innovation infrastructure and industrial clusters. The strength of these linkages depends on the organization of the university system as well as the funding mechanisms for new ventures. They determine the ability of a country to commercialize new ideas.

Measuring innovative capacity and technological specialization

There are several different indicators that can be used to measure changes in innovation, such as patenting activity, the share of high-technology sectors in total industry output, the proportion of firms introducing new or technologically improved products or processes on the market, etc.

Within the framework of the national innovative capacity they should be grouped into three sets of indices representing the three following areas: the common innovation infrastructure, the cluster-specific environment and linkages between these two (for the details see the annex table A).

The first group contains a broad characteristic of the national innovation system supplemented by some indicators selected to summarize:

- Stock of international patents (3 indicators)
- Human resources for innovation (5 indicators)
- Financial resources for R&D (1 indicator)

The second group of indicators reflects microeconomic perspective, in particular the effectiveness of industry clusters in creating and commercializing innovation. To this group belong 5 indicators describing innovation finance and 2 indicators of innovation output.

Furthermore, to this group can be added a broader characteristic of a country's technological specialization developed on the basis of patenting at the European Patent Office (EPO). To measure technological specialization of a country the RTCA (Revealed Technological Comparative Advantage) indicator is used. It is defined as follows:

$$RTCAi = Pij/\Sigma iPj / \Sigma jPij/\Sigma i\Sigma jPij,$$

where: Pij = number of patents in branch *j* of country *i*

ΣiPj = total number of world patents in branch *j*

ΣjPij = total number of patents of country *i*

ΣiΣjPij = total number of world patents

However, this indicator of innovativeness based on EPO patenting has some drawbacks. First of all, not all innovation can be patented. Moreover, patenting does not necessarily mean innovation. But on the other hand it reflects to some extent changes in technological performance at the sectorial level. The industrial sectors where the RTCA value is higher than 1 (RTCA>1) are relatively well developed in an analyzed country, while a value lower than one (RTCA<1) indicates that the sector does not for part of a country's technological specialization.

The third determinant of the national innovation capacity – the quality of linkages between the common innovation infrastructure and industry clusters – is measured here by 4 indicators referring to the transmission and application of knowledge. Complex innovations often depend on the ability to draw on diverse sources of information and knowledge, or to collaborate on the development of new knowledge. Therefore the transmission and application of knowledge is closely related to two indicators of innovation cooperation. One of these indicators is limited to small and medium size enterprises (SMEs), because almost all large firms are involved in cooperative projects. However, it captures only manufacturing firms that had any cooperation agreements on innovation activities with other enterprises or institutions. There is also another indicator of SMEs innovation activity, which measures a degree to which manufacturing SMEs that have introduced any new or significantly improved products or production processes have innovated in-house. And finally, the fourth indicator characterizing the quality of linkages between the common innovation infrastructure and industry clusters is home internet access. It measures the extent of the transmission of new knowledge (annex, Table A).

Most of the indicators are constructed as relative values (as a percentage of a country's GDP or per capita), so they allow a comparison of the performances of different countries. Data used for calculations are taken from the OECD, Commission of the European Communities (CEC) and the Polish Statistical Office (GUS) databases.

National innovative capacity: the case of Poland

Any national innovation system is characterized by a huge number of parameters. These include structural and economic characteristics as well as socio-institutional conditions. The most important factor that shapes national innovation system is industry-science relations. Interaction between the public research sector and industry takes various institutional forms. Measurable national differences include:

- Institutions that perform and fund R&D;
- Trends in R&D funding and performance patterns;
- Specialization in scientific disciplines.

According to these criteria there are four categories and ten sub-categories of national innovation systems of OECD countries (OECD, 2000b):

1. Countries with a very high share of government funding and performance:
   - University-based system (Turkey)
   - Broad-based system (Italy, Poland, Portugal)
   - Institute-based system (Hungary, Iceland)
2. Countries with a moderately high share of government funding and performance:
   - University-based system (Austria, Spain)
   - Broad-based system (France, Netherlands, Norway)
3. Countries with an average share of government funding and performance:
   - University-based system (Canada, UK)
   - Broad-based system (Denmark, Finland, Germany)
   - Institute-based system (the Czech Republic)
4. Countries with a low share of government funding and performance:
   - University-based system (Belgium, Ireland, Sweden, the USA)
   - Institute-based system (South Korea)

Poland falls into the category with a relatively high share of governmental funding and performance. Public expenditures on R&D constitute more than 60% of the total R&D expenditure and the greater part of research is performed by the public sector. On the other hand, however, there is a highly diversified system of public R&D organizations in Poland. These organizations include:

- universities;
- research units of the Polish Academy of Science, which concentrate mainly on basic research;
- branch research development units that focus more on applied research.

This diversity could be an advantage in the production of new knowledge on condition that more resources had been devoted to R&D activity. However, the transition of the socio-economic system in Poland has brought only slight changes in R&D financing. The state budget still remains the main source of financing for R&D activity and the ratio of public spending to GDP remains stable at a level of about 0.43%. It is significantly below the level of the leaders of EU15 and of that of leading accession counties, as well as below the EU average (table 1).

As far as human resources for R&D are concerned, Poland also performs below the EU average, but during the transition period it has been catching up. Since 1990 the number of new science and engineering graduates in Poland has increased considerably. The share of new S&E graduates in the 20-29 age class grew to 7.4%, and the growth trend of this indicator is positive and higher than the trend for the EU average. Moreover, the share of population with tertiary education also has been growing at a higher pace than that of the EU15 (CEC 2003). In 2002 the number of tertiary education students was about 3 times higher than in 1990 and it constituted 12.2% of the population of 25-64 years age. Poland's major strength relative to the EU15 and to the majority of acceding countries is employment in medium-high and high-tech manufacturing. This indicator of human resources is higher in Poland than the average in the EU, but still lower than that of Denmark, which is the EU leader in this respect (table 1).

The third group of indicators, which describes the common innovation infrastructure, is the stock of patents. In all three indicators of patenting activity Poland lags far behind the EU average and behind the level of the majority of acceding countries. In patent applications Slovenia is the leader among acceding countries, and Hungary occupies the second place. Poland's position has been weakening recently owing to the downward trend.

The characteristic of Poland's innovative capacity seen from a *cluster-specific perspective* is based on six indicators concerning innovation finance and output. For each of these indicators, Table 2 compares Polish results to that of the leading EU member state and to that of the leading acceding country. Poland is lagging in four out of six indicators. The largest gaps are in business expenditures on R&D, high-tech venture capital and early-stage venture capital. A matter of particular concern is that business R&D expenditures in both the EU15 and acceding countries averages are over 50 percent above Poland's indicator (table 2 and CEC, 2002).

However, in the remaining two indicators (i.e. innovation expenditures and sales of 'new to market' products) Poland performs above the EU average, but still below the leading countries. Nevertheless, the strongest point of Poland's innovative capacity is the growing effort to innovate in the manufacturing sector. In 2002 expenditure on innovation accounted for 4.1% of the total turnover in this sector compared to less than 0.5% in 1992. A similar growth trend has been observed in sales of 'new to the market' products. Its share in total sales of manufacturing increased from 5.4% in 1995 to 16.7% in 2002, and was higher than the average of the EU15 (GUS, 1996 and table 2).

Table 1: The common innovation infrastructure: Poland and other EU countries compared

Indicators	**Poland**	**EU mean***	**EU15 leader**	**Leader among acceding countries**
Stock of patents				
EPO patents applications (per million population)	2.5	161.1	Sweden (366.6)	Slovenia (40.7)
High-tech EPO patent applications (per million population)	0.2	31.6	Finland (136.1)	Slovenia (8.6)
USPTO patents granted (per million population)	1.1	80.1	Sweden (213.7)	Slovenia (13.1)
Human resources				
New S&E graduates (% of 20-29 years age class)	7.4	11.3	Ireland (21.7)	Lithuania (13.1)
Populations with tertiary education (% of 25-64 years age class)	12.2	21.5	Finland (32.4)	Lithuania (44.0)
Participation in life-long learning (% of 25-64 years age)	4.3	8.4	UK (22.3)	Slovakia (9.0)
Employment in medium-high and high-tech manufacturing (% of total workforce)	7.54	7.41	Denmark (11.36)	Slovenia (9.28)
Employment in high-tech services (% of total workforce)	N/A	3.57	Sweden (5.23)	Czech Rep. (3.09)
Financial resources				
Public R&D expenditures as % of GDP	0.43	0.69	Finland (1.02)	Slovenia (0.69)

* Weighted means based on summing the numerator and denominator across all analyzed countries (EU15 or acceding countries plus Bulgaria, Romania and Turkey).

Source: Based on: Commission of the European Communities (2003) *2003 European Innovation Scoreboard.* CEC Working paper SEC (2003) 1255, Brussels, p. 28

Table 2: Cluster-specific innovation environment: Poland and other EU countries compared

Indicators	Poland	EU15 mean*	EU15 leader	Leader among EU acceding countries
Innovation finance and output				
Business expenditures on R&D (% of GDP)	0.24	1.30	Sweden (3.31)	Slovenia (0.94)
Share of high-tech venture capital investment	17.5	45.4	Italy (71.2)	N/A
Share of early stage venture capital (% of GDP)	0.018	0.037	Sweden (0.098)	Czech Rep. (0.019)
Innovation expenditures (% of all turnover in manufacturing)	4.1	3.45	Sweden (6.42)	Slovakia (8.80)
ICT expenditures (% of GDP)	5.9	7.0	Sweden (9.8)	Estonia (9.6)
Sales of 'new to market' products (% of all turnover in manufacturing)	16.7	10.5	Finland (27.2)	Malta (37.8)

* Weighted means based on summing the numerator and denominator across all analyzed countries (EU15 or new EU members plus Bulgaria, Romania and Turkey).

Source: Based on:

1/ Commission of the European Communities (2003) *2003 European Innovation Scoreboard*. CEC Working paper SEC (2003) 1255, Brussels, p. 28

2/ GUS (2003), *Rocznik statystyczny 2003*, Warszawa, GUS, p. 340.

3/ Commission of the European Communities (2002) *2002 European Innovation Scoreboard, Technical Paper No 2, Candidate Countries*. CEC, Working Paper SEC (2002) 1349, Brussels 2002, p. 11.

To have a complete picture of Poland's technological development it is worth assessing Poland's technological specialization. Evaluating the revealed technological comparative advantage (RTCA) indicator for different branches of the Polish economy allows us to define the sectors where Poland has a comparative technological advantage. Table 3 shows the results of calculations. According to these results, Poland has a comparative technological advantage in medium-high and medium-low technology industries for the whole period analyzed.

In medium-low industries, Poland's comparative technological advantage appeared continuously in metallurgy and in some other medium-low technology manufacturing (such as some sport equipment, simple machine parts, etc.).

Poland's technological specialization in medium-high technological industries appeared in 1992. It was mainly a result of patenting in the motor vehicle industry and other transportation equipment industries. In some separate years the RTCA indicator also was higher than 1 in chemical and machinery industries, but it was rather an occasional feature than a constant trend.

Moreover, it should also be pointed out that there are some occasional shifts towards technological comparative advantage in low-tech industries, such as wood processing and textiles. However, the RTCA index is lower than 1 in all branches belonging to high-technology industries, which indicates that they are comparatively technologically disadvantageous for Poland.

Table 3: Poland's comparative advantages in technology, 1991-2000

Industry branches (see Annex table B)	**RTCA indicator**									
	1991	**1992**	**1993**	**1994**	**1995**	**1996**	**1997**	**1998**	**1999**	**2000**
High-technology industries	*0.296*	*0.608*	*0.883*	*0.443*	*0.724*	*0.415*	*0.263*	*0.162*	*0.611*	*0.345*
Aircraft	0.000	0.000	0.000	0.000	0.000	0.000	0.000	0.000	28.006	0.000
Office and computing equipment	0.000	0.000	0.000	0.000	0.968	0.000	0.000	0.000	0.000	0.292
Drugs and medicines	0.511	0.900	1.226	0.650	0.856	0.679	0.470	0.151	0.786	0.449
Radio, TV and communication equipment	0.000	0.000	0.000	0.000	0.000	0.000	0.000	0.290	0.000	0.245
Medium-high-technology industries	*0.851*	*1.691*	*1.007*	*1.239*	*1.629*	*1.298*	*1.256*	*1.520*	*1.168*	*1.401*
Motor vehicles	0.000	4.484	0.000	2.253	2.815	2.634	1.540	4.310	3.095	10.291
Chemicals excl. drugs	1.813	1.582	1.783	1.622	1.271	0.338	0.644	1.200	1.695	0.478
Other transport equipment	0.000	0.000	0.000	0.000	1.626	0.000	1.787	3.265	1.814	1.399
Machinery and professional goods	0.368	1.818	0.657	1.057	1.822	2.479	1.696	1.092	0.725	1.150
Medium-low-technology industries	*2.329*	*0.988*	*1.106*	*1.494*	*1.134*	*1.022*	*1.475*	*1.672*	*1.921*	*1.349*
Rubber and plastic products	0.000	0.000	0.000	0.000	0.000	0.000	0.000	0.000	0.000	0.000
Other manufacturing	1.994	1.462	1.067	0.607	0.000	1.564	3.905	2.605	2.914	1.141
Metals and products	3.246	1.065	0.782	2.243	1.714	0.516	0.343	1.676	1.315	0.745
Non-metallic mineral products	0.000	0.000	0.000	2.298	0.000	2.529	0.000	0.000	3.185	1.246
Petroleum refineries	0.000	0.000	12.122	0.000	9.666	0.000	0.000	0.000	0.000	4.075
Low-technology industries	*0.000*	*0.000*	*0.976*	*0.565*	*0.000*	*1.198*	*0.972*	*0.511*	*0.336*	*1.005*
Wood products and furniture; paper, pulp and printing	0.000	0.000	0.000	1.059	0.000	2.334	2.186	0.531	0.000	2.718
Textiles, apparel and leather	0.000	0.000	0.000	0.000	0.000	0.000	1.796	0.000	0.000	1.300
Food, beverages and tobacco	0.000	0.000	1.577	0.446	0.000	0.947	0.311	0.584	0.513	0.374

The RTCA indicator is defined by equation 1. When RTCA>1 a country enjoys comparative technological advantage. RTCA<1 shows that the industry is technologically disadvantaged.

Source: Author's calculations based on Polish Patent Office data.

The change in technological specialization during 1991-2000 can be measured by the correlation between RTCAs. The high value of the correlation coefficient shows that the technological specialization is relatively stable, while its low value indicates considerable changes in technological advantages.

In Poland, the correlation coefficient between RTCA in 1991 and 2000 is negative and relatively low (-0.15194). This means that over the period analyzed, there were some significant changes in Poland's technological specialization (table 4). The correlation coefficient between RTCA in 1991 and 1994 is also negative and low (-0.009417), while it is positive and relatively higher (0.5182) between 1994 and 2000. This statistical evidence suggests that shifts in technological specialization were more likely in 1991-1994 than in 1995-2000. It seems that these changes in Poland's technological specialization were caused by factors related to the transition of the socio-economic system.

Table 4: Changes in Poland's technological comparative advantage 1991-2000

Variables correlated	RTCA 1991 and RTCA 2000	RTCA 1991 and RTCA 1994	RTCA 1994 and RTCA 2000	RTCA 1998 and RTCA 2000
Pearson's correlation coefficient	-0.15194	-0.009417	0.5182	0.433282

Source: Calculations based on table 3.

Finally, another determinant of Poland's national innovative capacity is *the quality of linkages* between the common innovation infrastructure and industrial clusters, reflected in 4 indicators concerning transmission and application of new knowledge.

A comparison of Poland with EU members and acceding countries shows the relatively weak position of Poland's economy in this respect (table 5). Only one out of four indicators is above EU average. The weakest point of the system of transmission and application of knowledge in Poland is the low level of innovativeness within SMEs. The share of SMEs innovating in-house in all SMEs in manufacturing is much lower not only in comparison with the EU average, but also when compared with the results of other acceding countries. The leader among acceding countries is Estonia, where this share is nearly ten times higher than that of Poland (table 5).

There is also much to be done in Poland in the field of home Internet access. The development of Internet connections is closely linked to the fall in the costs of ICT. Competition in the local telecommunication loop is of particular importance in this respect. In Poland there is one dominant operator – Telekomunikacja Polska SA. In the past it was a state-owned enterprise, but its privatization in 2000 did not bring changes in its dominant position on the Polish market, as it was not accompanied by necessary regulatory reform in the telecommunication sector. Therefore, the absence of competition in the local telecommunication loop along with the growing demand for Internet access enabled the dominant operator to keep prices high for telecommunication services. In Poland the prices of Internet connection are the highest in the region. As a result Poland, similarly to Lithuania and Latvia, performs the worst among EU acceding countries in respect of Internet access and use.

EU accession will hopefully change the situation on Polish telecommunication market. On the one hand EU accession brings the increase of VAT tax on Internet services in Poland from today's 7% to 22%, which may limit the use of the Internet. On the other hand, however, as a result of EU accession Polish telecommunication market will be liberalized. Competition in the telecommunication sector will force operators to lower prices, which may increase Internet use.

Table 5: Quality of linkages between the common innovation infrastructure and industrial clusters: Poland and the EU compared

Indicators	Poland	EU15 mean*	EU15 leader	Leader among EU acceding countries
Transmission and application of knowledge				
Firms involved in innovation cooperation (% of all innovating firms)	29.0	27.0	Finland (38.0)	N/A
SMEs involved in innovation cooperation (% of manufacturing SMEs)	6.9	9.4	Finland (22.0)	Lithuania (12.1)
SMEs innovating in-house (% of manufacturing SMEs)	4.1	37.4	Denmark (55.1)	Estonia (39.1)
Home Internet access (% of all households)	0.08	0.51	Sweden (0.97)	Malta (0.44)

* Weighted means based on summing the numerator and denominator across all analyzed countries (EU15 or new EU members plus Bulgaria, Romania and Turkey).

Source: Based on:

1/ Commission of the European Communities (2002) *2002 European Innovation Scoreboard, Technical Paper No 2, Candidate Countries*. CEC, Working Paper SEC(2002) 1349, Brussels 2002, p. 11;

2/ GUS (2001) Działalność innowacyjna przedsiębiorstw przemysłowych w latach 1998-2000. GUS Warszawa, pp. 86-87;

3/ Commission of the European Communities (2003) *2003 European Innovation Scoreboard*. CEC Working paper SEC (2003) 1255, Brussels, p. 28.

Conclusions

Innovation and technical change appear of great importance for strengthening growth performance; therefore it is essential to analyze the determinants of the national innovative capacity. The concept of the national innovative capacity, developed by Porter, Furman and Stern allows us to assess the performance of Poland and compare it with the EU average, and with the performance of other EU acceding countries. In particular, this paper explores the Poland/EU gap for a set of innovative capacity indicators.

Within the group of indicators that characterize the common innovative infrastructure there is only one area where this gap has been closed, i.e. employment in high-tech manufacturing. However, a catching-up process can be observed in the development of human resources, i.e. S&E graduates and tertiary education. Some signs of recovery can also be observed in patenting activity of Polish citizens in the European Patenting Office, but the gap between Poland and the EU still remains huge. On the other hand, the gap is widening in high-tech patenting, as well as in public R&D expenditures.

With regard to the cluster-specific innovation environment, an increasing gap between Poland and the EU average is in business R&D. The only advance of Poland compared to the EU average is in innovation expenditures and sales in manufacturing sector. Moreover, a positive shift in Poland's technological specialization has been observed since 1991: from low-tech and medium-low-tech industries to some medium-high-tech industries, in particular from food and metallurgy manufacture to the manufacturing of motor vehicles and chemicals. And a long-lasting catching-up process exists in ICT expenditures, where the gap between Poland and EU average has been cut since 1995.

The analysis of indicators that characterize the quality of linkages between the common innovation infrastructure and industrial clusters reveals the weakest point of the system of transmission and application of knowledge in Poland. It is a low level of innovativeness within SMEs. Regarding SMEs innovating in-house, Poland ranks lower than the EU average and the majority of EU acceding countries.

Poland's position in relation to the EU members and acceding countries can be summarized using the synthesis indicator, i.e. Summary Innovation Index (SII-2), based on the twelve indicators of innovative capacity (for details of the calculation of SII-2 see: CEC, 2003).

In terms of the SII values, countries can be divided into 4 groups: countries that are moving ahead, losing momentum, catching up or falling further behind.

Poland falls into the 'catching up' group. To this group belong also other acceding countries, and some EU members, such as Ireland, Luxembourg, Austria, Spain, Portugal and Greece and the EU candidate countries: Romania and Turkey. The 'moving ahead' group consists of the innovation leaders: Sweden, Finland, the UK, Denmark and Belgium. There are three 'losing momentum' European countries: the Netherlands, Germany and France. And countries that fall further behind are Italy and Bulgaria (CEC, 2003).

All acceding countries fall into the 'catching up' group, which reflects the strong growth of their innovative capacity indicators. However, a large part of this growth is due to the fact that these indicators were at a very low level in the beginning of the 1990s, when transition of socio-economic systems (including science and technology) began. In the case of Poland a problem of further development of the national innovative capacity concerns the falling public and business R&D expenditures. It is a serious handicap to the improvement of Poland's innovation performance. Therefore, in the case of Poland, the catching-up trend may not be sustainable in the near future.

References

Aghnion, P. Howitt, P. (1998) Endogenous Growth Theory. MIT Press, Cambridge MA.

Archibugi, D., Michie, J. (eds) (1998) Trade, Growth and Technical Change. Cambridge, Cambridge University Press.

Bell, M., Pavitt, K. (1993) Technological Accumulation and Industrial Growth: Contrasts between Developed and Developing Countries. "Industrial and Corporate Change", Vol. 2, No. 2, 1993.

Commission of the European Communities (2001) Statistics on Innovation in Europe, EUROSTAT, CEC, Luxembourg.

Commission of the European Communities (2002) 2002 European Innovation Scoreboard, Technical Paper No 2, Candidate Countries, CEC, Working Paper SEC (2002) 1349, Brussels.

Commission of the European Communities (2003) 2003 European Innovation Scoreboard. CEC, Working paper SEC (2003) 1255, Brussels.

GUS (1996) Rocznik statystyczny 1996, Warszawa, GUS.

GUS (2000) Raport o stanie nauki i techniki w Polsce. GUS, Warszawa.

GUS (2001) Działalność innowacyjna przedsiębiorstw przemysłowych w latach 1998-2000. GUS Warszawa

GUS (2003), Rocznik statystyczny 2003, Warszawa, GUS.

Metcalfe, S. (1995) The Economic Foundations of Technology Policy: Equilibrium and Evolutionary Perspectives. In: P. Stoneman (ed.) Handbook of the Economics of Innovation and Technical Change. Blackwell, London.

Nelson, R., Rosenberg, N. (1993) National Innovation System. Oxford University Press, Oxford.

OECD (1994) Patent Manual 1994. Using Patent Data as Science and Technology Indicators. OECD, Paris.

OECD (1995) Classification of High-Technology Products and Industries. DSTI/EAS/IND/STP(95)1. OECD, Paris.

OECD (1999) OECD Science, Technology and Industry Scoreboard 1999. OECD Paris.

OECD (2000a) Main Science and Technology Indicators 2000/1. OECD, Paris.
OECD (2000b) Science, Technology and Industry Outlook. OECD, Paris.
Pavitt, K. (1984) Sectoral Patterns of Technical Change: Towards a Taxonomy and a Theory. "Research Policy", Vol. 13, 1984.
Porter, M. (1990) The Competitive Advantage of Nations. The Free Press, New York.
Romer, P.M. (1989) Endogenous Technical Change. NBER Working Paper No. 3210, Cambridge Massachusetts 1989.
Stern, S., Porter, M., Furman, J.L. (2000) The Determinants of National Innovative Capacity. National Bureau of Economic Research, Working Paper No. 7876, September 2000.
World Intellectual Property Organization (1989) International Patent Classification. Carl Haymanns Varlag KG, Munich.
Weresa, M.A. (2002), Wpływ handlu zagranicznego i inwestycji bezpośrednich na innowacyjność polskiej gospodarki. "Monografie i Opracowania" No. 504, SGH Warszawa.

Annexes

Table A: Innovative Capacity Indicators

Area of interest	Indicator
The Common Innovation Infrastructure	
The model of national innovation system (NIS) (2 indicators)	1. A share of government funding in R&D 2. Major institutions that perform R&D
Stock of patents (3 indicators)	1. EPO patent applications (per million population) 2. EPO high-tech patent applications (per million populations) 3. USPTO patents granted (per million populations)
Human resources for innovation (5 indicators)	1. New S&E graduates (‰ of 20-29 years age class) 2. Populations with tertiary education (% of 25-64 years age class) 3. Participation in life-long learning (% of 25-64 years age) 4. Employment in medium-high and high-tech manufacturing (% of total workforce) 5. Employment in high-tech services (% of total workforce)
Financial resources for R&D (1 indicator)	1. Public R&D expenditures as a% of GDP
Cluster-Specific Innovation Environment	
Innovation finance and output (6 indicators)	1. Business expenditure on R&D (% of GDP) 2. Share of high-technology venture capital investment 3. Share of early stage venture capital in GDP 4. Innovation expenditure (% of all turnover in manufacturing) 5. ICT expenditure (% of GDP) 6. Sales of 'new to market' products (% of all turnover in manufacturing)
Technological specialization (1 indicator)	1. Revealed technological comparative advantage (RTCA)
Quality of Linkages between the common Innovation Infrastructure and Industrial clusters	
Transmission and application of knowledge (4 indicators)	1. Firms involved in innovation cooperation (% of all innovating firms) 2. SMEs involved in innovation cooperation (% of manufacturing SMEs) 3. SMEs innovating in-house (% of manufacturing SMEs) 4. Home Internet access (% of all households)

Source: Based on OECD and EC innovation indicators.

Table B: List of industries

Industries	NACE Rev.1	International patent classification code
High-technology industries		
Aircraft	35.30	B64 B82
Office and computing equipment	30	B81 G06 G11
Drugs and medicines	32	H04
Radio, TV and communication equipment	32	H04
Medium-high-technology industries		
Motor vehicles	34	B60 B62 F15
Chemicals excl. drugs	24 excl. 24.4	C01 C05- C25 C02 C09 C30
Other transport equipment	35.2, 35.44 35.5	B61 B65 B66
Machinery and professional goods	29 31 33	B23 G01- G12 F01- G05 H01- F04 G07- H03 G09 H05
Medium-low-technology industries		
Rubber and plastic products	25	B29
Other manufacturing	35.1 36.4 36.2 36.5 36.3 36.6	A46 B63 F17 A62 B67 F21 A63 D01 F22 B43 E01- G09 B44 E04 G10
Metals and products	28 27.3 27.1 27.4 27.2 27.5	B04 B26 E05 B05 B30 F16 B21 C21 F23 B22 C22 F24 B25 C23 F27
Non-metallic mineral products	26	B02 B28 C04 B03 C03 E21
Petroleum refineries	23	C10
Low-technology industries		
Wood products and furniture; paper, pulp and printing	20 36.1 21 22	A47 B41 B27 B42 B31 D21 B32
Textiles, apparel and leather	17 19 18	A41-A45 C14 B68 D02-D07
Food, beverages and tobacco	15 16	A01 C11-C13 A21-A24

Source: Author's classification based on:

1/ OECD (1999) OECD Science, Technology and Industry Scoreboard 1999. OECD Paris, p.106.

2/ World Intellectual Property Organization (1989) *International Patent Classification.* Carl Haymanns Varlag KG, Munchen.

3/ OECD (1994) Patent Manual 1994. Using Patent Data as Science and Technology Indicators. OECD, Paris, pp. 63-94.

Chapter 10
Institutional and Financial Mechanisms in the Development of an Innovative Economy in the Saratov region, Russia

V. Atoyan, Yu. Tchebotarevsky, A. Slepoukhin, N. Kazakova, V. Tiourina
Saratov State Technical University
Politechnicheskaya Street 77
Saratov, 410054
Russia

Summary

In this chapter, the processes of formation and development of institutional and financial mechanisms in the development of a regional innovation system in the Saratov region are described.

Introduction

The beginning of the third millennium has obviously confirmed the fact that science and technical progress have turned into a main source of economic growth. Permanent innovation activity has become a basis of modern economic progress. The competition for new ideas and their fastest implementation will continue to increase around the world. Under these circumstances, institutional and financial provision of new knowledge production and utilization has critical importance as a key factor of national competitiveness.

All advanced and rapidly developing countries realized this situation a long time ago. Implementing active, purposeful innovation policy is one of the important activities both for federal authorities and for regional ones.

Intensification of state innovation policy in the Russian Federation

Supporting national competitiveness, not only at the expense of exploiting limited natural resources but also through the effective utilization of intellectual and innovation potential, is an acknowledged aim of the Russian Federation as well. Russian legislative and executive authorities have been rethinking the role of innovation activity in socio-economic development, as is confirmed by the intensification of development of the innovation activity normative base (see table 1).

In November 2001, the Science and High Technology Council was created under the President of the Russian Federation. The functions of this advisory body are to inform the President about the status of R&D policy, to support his cooperation with research organizations, and to suggest the priorities of this policy.[1] In 2002, the President and the Government of Russia approved the first fundamental document "Principles of R&D Policy in Russian Federation up to 2001 and Further Perspectives"[2].

In February 2004 the next step on Russia's transition to innovative development was made. The document "Principles of Russian Federation Policy in the field of national innovation system (NIS) development up to 2010" was approved. This document determines the goals, tasks, and key directions of state policy on national innovation system development, mechanisms and measures of this policy, as well as predicted indicators of NIS development. It is noted that this policy is a part of state socio-economic and scientific-technical policy connecting closely with other state policies (budgetary, custom, credit and tax policies). It is oriented towards creating an economy that is based on new knowledge and progressive technologies as well as developing high-tech production of competitive products.

Table 1: Legislative regulation of innovation activity in Russian Federation

Dates	**Documents**
December, 1995	Government Decree (Dec., 26, 1995, Russian Federation) "The Most Important Measures in the Development and Support of Innovative Activities in Industry"
1996	Russian Federation Act "On Science and Technological Politics"
July, 1998	Government Decree (July, 24, 1998, Russian Federation) "The Concept of Innovative Politics in Russian Federation"
November, 2001	President of the Russian Federation Decree (8.11.2001 № 1301) "On Science and High Technology Council of Russian Federation"
March, 2002	Principles of R&D Policy in the Russian Federation up to 2010 and Further Perspectives
February, 2004	Principles of Russian Federation Policy in the Field of the National Innovation System Development up to 2010

In the accepted document the national innovation system (NIS) as an economic system is considered to be a complex of economic agents that interact within the processes of production, implementation, dissemination and utilization of profitable knowledge and technology. The main elements of the Russian Federation NIS are the following:

- a subsystem of knowledge production based on fundamental research;
- a subsystem of professional education;
- high-technology production;
- NIS infrastructure.

State policy on national innovation system development is conducted along the following lines:

- an improvement of the normative base which is providing the commercialization of intellectual activity resulting in:
  - improvement of the normative base of protection and use of the intellectual activity results produced at the expense of budgetary funds;

1 "О Совете при Президенте Российской Федерации по науке и высоким технологиям". Указ Президента Российской Федерации от 8.11.2001. № 1301.

2 Основы политики Российской Федерации в области развития науки и технологий на период до 2010 года и дальнейшую перспективу // Поиск. № 16 (674). 19 апреля 2002 г. С.8-10.

- working-out of the normative base for accelerating NIS infrastructure development;
- formation of a system of legislative and normative acts above all on taxation, custom, tariff and technical regulation;
- working out normative regulations for the use of the scientific-technical and military-technical information system for information support of innovation activity.

NIS infrastructure development for assistance to transfer of commercially effective R&D results to businesses, pre-production and entering the market of innovation products. This will be carried out:

- with state support and shared participation, by executive authorities, local self-governments and businesses on the territories which have highly concentrated innovation potential including technopolises, science towns, academic towns, and closed administrative-territorial units;
- by the integration of science, education and industry, and also by accelerating knowledge and technology transfer between the defence industry and civil economy;
- an improvement in the state support of domestic innovation projects implementation:
- through the coordination of federal, regional and industrial innovation projects to consolidate and concentrate budget and off-budget funds on selected priorities;
- through the movement to the complex solving of large-scale problems of innovation development of regions and high-tech industries in the framework of Russian Federation (RF). Priorities of science and technology development and RF Register of critical technologies.

However, it should be noted that these processes, which are now recognized on government level, began much earlier. From the middle of 1990 they were already being implemented in some regions of the Russian Federation including the Saratov region.

Brief description of the Saratov region

The Saratov region is one of the key regions in Russia. On the one hand, the region is one of central territories within the European part of Russia and the Commonwealth of Independent States (CIS). On the other hand, it borders another state (Kazakhstan). The region is well located in the centre of a large grain-producing zone on the crossroads of trade and transport routes. The best way to Middle Asia and the Kazakhstan Republic is through the Saratov region.

The Saratov region is a big industrial centre in Russia. Electrical power engineering, mechanical engineering, the chemical and petrochemical industries and the food industry have the largest specific gravity in the sectoral make-up of regional industry (Fig.1).

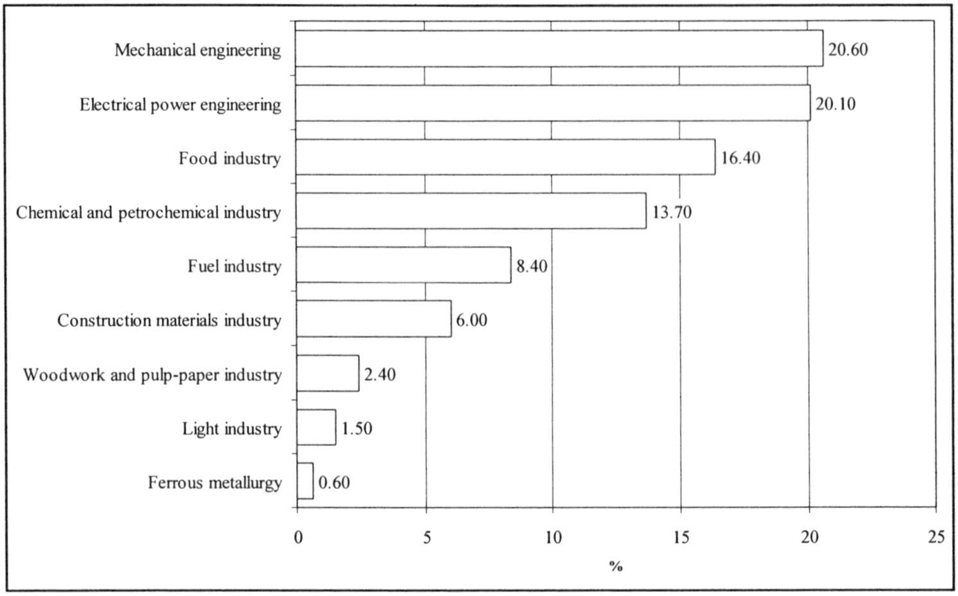

Fig. 1: Sectoral make-up of Saratov region industry[3]

Despite the crisis of the 1990s, Saratov has continued to be a large scientific-technical and educational center. R&D is carried out by 8 high schools, 18 industrial research institutes, 10 academic institutes of the Russian Academy of Sciences and 27 constructing organizations and research laboratories at enterprises, in which approximately 1,000 professors, 3,500 doctors in science and more than 10,000 highly qualified specialists are employees. In 2001, the number of R&D personnel grew by 2.5% in comparison with 2000. Total volume of R&D equals more than 1.2 billion roubles in 2001 and grew by 40% in comparison with 2000. R&D is carried out in such high-tech industries as mechanical engineering, electronic and aerospace industry, ITT (more than 50% of the total volume), chemical industry (about 35% of the total volume) and the fuel and power industry (about 15% of the total volume).

Everything mentioned is reflected in the positive dynamics of basic regional indicators in 2001-2002. The optimistic fact is that production growth was achieved mostly at the expense of using regional innovation potential. Since 1997 the number of innovating enterprises has been increasing in the region. Most of this is implementing innovations in the expansion of the range of goods, in the maintenance and development of markets, and in the improvement of labour conditions. Most innovating enterprises are in the instrument industry (including specialized instrument-making plants of the defence industry), mechanical engineering and electrical power engineering.

At the present time in the Saratov region all three lines of state policy on the national innovation system development are being fulfilled more or less actively – an improvement of the normative base which is providing the commercialization of the results of intellectual activity, NIS infrastructure development and an improvement of state support of domestic innovation projects implementation.

[3] Стратегический план развития Саратовской области. – Саратов: 2000. – С.5.

Improvement of the normative base of innovation activity in Saratov region

The experience of developed countries shows that it is impossible to intensify innovation activity without adequate legislative support in this area. Unfortunately, in Russia there is no Federal Act which is able to control the innovation activity in the country as a whole. But, this negative factor is partly compensated on the regional level – more than 30 regions have local acts to control research, technology and innovation activities.

Existing legislation of the Saratov region in the innovation sphere is based on Russian Federation legislation and consists of the Saratov region Act "On Innovations and Innovation Activity", the projected new Saratov region Act "On support of specialized agents of innovation activity " and other regional normative rules.

Saratov region Act "On Innovations and Innovation Activity"[4]

The first regional act in the Russian Federation "On Innovations and Innovation Activity" was approved in the Saratov region in 1997 on the initiative of Saratov State Technical University scientists and appears to be the foundation for passing analogous acts in other regions of the Russian Federation.

Over the last decades great changes have taken place both in the country and the Saratov region. Because of these changes, it was necessary to strengthen innovation activity and its state supports, and to renew the definition of innovation agents, regulation mechanisms, etc.

So at the beginning of 2003, the Saratov Region Duma adopted a new variant of the act "On innovation and innovation activity" clarifying these concepts. Fortunately, this variant was initiated not by researchers but by the regional government, showing that they understand the importance of innovations in the new economics.

In this Act the following definitions are used:

- *innovation* – a result of science and scientific-technical activity accepted as an object of intellectual property according to existing legislation;
- *innovation activity* – an activity aimed at the implementation of innovations, which are results of science and scientific-technical activity, and including both the utilization of innovations and assistance in this process. The activity of natural and legal persons is to be considered as innovation activity conditional on the use of an officially protected object of intellectual property within this activity.
- *innovation infrastructure* – a complex of natural and legal persons providing effective functioning innovation subjects.

An important characteristic of this Act is the normative fixing interconnection between innovation activity and intellectual property. In the theoretical aspect this interconnection reflects the transformation of scientific knowledge into a real market product, and raising the economic nature of knowledge. In the practical aspect it gives a possibility to commercialize the results of intellectual activity on the basis of their being legally protected and to identify clearly the subjects who innovate actively and have a right to receive support from the government.

In this Act there are the following terms within the Saratov region:

- innovation activity subjects;
- innovation activity results (products);

---

[4] Закон Саратовской области „Об инновациях и инновационной деятельности" (новая редакция). Принят Саратовской областной Думой в марте 2003 г.

- forms of innovation activity being carried on and its specifics;
- the status of the innovation programme and innovation project;
- innovation management within the Saratov region;
- regional government authority in innovation management;
- innovation activity of foreign legal and natural persons;
- financing innovation activity;
- stimulating innovation activity;
- rights of innovation subjects.

Saratov region Act "On support of specialized subjects of innovation activity"

Recently the scientists of Saratov universities (mainly Saratov State Technical University) have been working diligently on the project of the new Saratov region Act entitled "On support of specialized subjects of innovation activity ". It has been developed under the first act, and contains more specific regulating mechanisms of the state support given to innovation process participants. These mechanisms include financial and organizational techniques, tax concessions, intellectual property protection, etc.

This Act regulates interactions between regional authorities, specialized subjects of innovation activity, and users of innovations. It is aimed at creating conditions for state regional innovation policy, for state regulation and state support of innovation activity within the Saratov region, and the effective use of regional innovation potential.

In this Act the following definitions are used:

State regional innovation policy – complex of goals and priorities of innovation activity development within the region, ways and means of achieving them, creating conditions for a successful functioning regional innovation system, and accelerated transformation of knowledge into modern technologies and products on the base of interacting federal and regional authorities.

State support of innovation activity within the region – a system of measures, aimed at the formation and development of an innovation products market and creating favourable conditions for innovation activity within the region.

Regional innovation system – a complex of organizations and enterprises, as well rules and mechanisms, which initiate, create and disseminate innovation products and technologies.

Innovational–Technological Centre (ITC) – specialized subject of innovation activity, jointly established by enterprise and research or an educational organization to implement innovation projects to produce high-tech products.

Innovational–Industrial Complex (IIC) – specialized subject of innovation activity, jointly established by one or a few enterprises and research and education organizations to implement innovation programmes.

Technopark – specialized subject of innovation activity, established usually near or in the structure of a university or another research organization to carry out the small-scale production of high-tech products.

Small innovation enterprise – specialized subject of innovation activity, established by enterprises, organizations or persons to carry out the small-scale production of high-tech products.

Specialized subject of innovation activity – a legal entity or an entity having legal entity authority by proxy according to RF legislation, which carries out the science and/or innovation activity as a primary activity and acts according to constituent documents of a specialized subject of innovation activity.

According to the present Act specialized subjects of innovation activity (SSIA) are as follows:

innovational–technological centres ;

innovational–industrial complexes;

technoparks and incorporated firms;

small innovation enterprises;

organizations and enterprises of innovation infrastructure.

For supporting innovation activity within the region, the regional authorities fulfil the *following measures*:

administrative and management measures (legal acts regulating a regional innovation system, normative documents, strategic programs etc.);

- economic measures:
- tax policy;
- credit policy;
- direct financing of selected innovation projects and programs, specialized subjects of innovation activity and innovation infrastructure from the regional budget;
- real estate policy (allocation of accommodation and land for the setting up of specialized subjects of innovation activity);
- customs and tariffs policy;
- innovation programmes and projects of selected industries and enterprises.

For organizational provision of the measures presented in this Act, the Governor of the Saratov region appoints an authorized agency under the Saratov regional government which is responsible for regional innovation policy fulfillment. The work of this agency is aimed at the development of the regional innovation system.

Public accreditation of specialized subjects of innovation activity. To identify clearly the subjects who innovate actively and have a right to receive support from government, this authorized agency carries out the public accreditation of specialized subjects of innovation activity and issues them a certificate of public accreditation. If the scope of innovation activity of the economic subject is not less than 70% of the total scope of its activity, it obtains the certificate of public accreditation as a specialized subject of innovation activity. This certificate is a basis for granting government support and tax preferences, determined in the present Act, and other preferences for innovators according to RF legislation.

Regional government support of specialized subjects of innovation activity and innovation infrastructure is fulfilled by the the authorized agency under the Saratov regional government. Regional government support of innovation activity is conducted along the following lines:

1. Support of the specialized subjects of innovation activity in the sphere of *intellectual property protection*. It works through establishment of specialized offices assisting for specialized subjects of innovation activity and individual patent holders in patenting, licensing and patent protection, foreign patenting, seeking partners for licensee agreements, consulting etc.
2. *Material and technical support* of specialized subjects of innovation activity: preferential assignment of mismanaged property and bankrupt's assets, providing pre-emptive buyout of leased property.
3. Regional government support by *price regulation* is supposed to be through controlling the price level of resources which are necessary for innovation activity. This is achieved

by contract prices and preferential pricing on local fuel, energy and other resources and services.

4. Support of specialized subjects of innovation activity by *leasing* through:
   - an attraction of domestic and foreign leasing companies to innovation projects implementation as lessors or guarantors;
   - creating leasing companies (interregional, regional, corporative companies, with participation of regional authorities etc.) to lease modern high-tech equipment;
   - attracting banks for crediting of leasing for innovation project implementation;
   - granting of additional benefits and guarantees by regional authorities for leasing;
   - developing insurance of leasing transactions of specialized subjects of innovation activity;
   - refunding part of lease payments to specialized subjects of innovation activity from the regional development budget and other sources.
5. *Insurance* of specialized subjects of innovation activity will be an insurance of specialized subjects of innovation activity themselves and an insurance of their actions. It is designed to develop an insurance of specialized subjects of innovation activity by:
   - the creation of a specialized insurance company for the insurance of specialized subjects of innovation activity;
   - the participation of specialized subjects of innovation activity in the creation of insurance corporations.
6. *Personnel coverage* of specialized subjects of innovation activity is to be conducted in the following directions:
   - short-term and long-term courses of training/retraining personnel in innovation management;
   - additional high professional education in innovation management on the basis of the Saratov subsidiary of the Russian State University of Innovation Technologies and Entrepreneurship;
   - conducting lectures, workshops, training sessions in innovation management;
   - in-plant training of Saratov university students on the basis of specialized subjects of innovation activity;
   - consulting in innovation management;
   - opening branches of university departments on the basis of specialized subjects of innovation activity;
   - opening recruiting agencies for the innovation sphere;
   - and other directions providing qualified personnel coverage of specialized subjects of innovation activity and effectively using human potential in the region.
7. *Financing* innovation programmes and projects of specialized subjects of innovation activity and innovation infrastructure using:

- specifically targeted funds from regional and local budgets;
- funds from the development budget of the Saratov region;
- joint investments of banks and the development budget;
- Venture Innovation Fund – non-commercial organization;
- own funds of specialized subjects of innovation activity and other sources.

The development budget is supposed to be created annually in the structure of the Saratov region budget, according to the regional budget low. Priority financing innovation programs and projects (not less than 50%) should be placed within the development budget. To attract commercial banks to invest in innovations, joint financing of specialized subjects of innovation activity by banks, the development budget and the own funds of specialized subjects of innovation activity should be used.

The regional government and businesses will create an off-budgetary Venture Innovation Fund for the Saratov region. The capital for this Fund is generated by voluntary deductions by enterprises, organizations and the people. The sum of deductions is included in product cost (but not more than 1.5% from cost).

The Venture Innovation Fund activities are as follows:

- financing innovation programs and projects of specialized subjects of innovation activity;
- free-of-charge allocation of preferential credits, interest-free loans and financial assistance for specialized subjects of innovation activity.

The Venture Innovation Fund gives money for innovation programs and projects of specialized subjects of innovation activity by making contracts according to the Statute of Fund. To stimulate innovation activity in Saratov region, specialized subjects of innovation activity and innovation infrastructure obtain tax incentives as follows:

1. profits tax relief for Venture Innovation Fund under condition of reinvesting profit in innovation activity;
2. earned income tax relief for specialized subjects of innovation activity if this income is a result of innovation programs and projects;
3. enterprise assets tax relief for specialized subjects of innovation activity during their participation in innovation programs and projects;
4. local tax exemption for specialized subjects of innovation activity during their participation in innovation programs and projects.

The projected new Saratov region Act "On support of specialized subjects of innovation activity " has already been discussed at a session of the Saratov region Duma.

Strategic Plan of Saratov region development

Both of the above-mentioned acts are the most highly regarded with reference to the innovation sphere.However, in modern society there is an important task of identifying priorities and coordinating the actions of all economic actors, providing interconnection and interaction of all legislative and normative acts, and developing mechanisms for integrating several programs into an integral system. To solve this task, the Strategic Plan of the Saratov region development was adopted in 2000.

This document formulates the strategic directions of regional development, goals, measures for achieving them, and identifies mechanisms of control by authorities, business and people. The general goal of the strategy of Saratov region development is an improvement in the quality of life of the population. A formation of preconditions of a post-industrial economy is noted as the main way of achieving this. That is why a strategy of the increasing competitiveness of the region, enterprises, products and services of Saratov producers is called one of seven main priorities of regional development. Among the paths leading to fulfilment of this strategy are: a) the creation of a highly effective innovation-production system, that proposes a reconstruction of all industries on a new technological base and their integration into a complete system of development, production and distribution of competitive high-tech products; b) technological reconstruction of the regional economy on the basis of ITT, that proposes the support of Internet-providers and producers of electronics and the creation of an integral regional information system.

The Strategic Plan should solve the task of the integration of efforts and resources of all economic actors and, finally, carry out a main strategy of worldwide development within Saratov region.

The development and practical realization of these acts will favour the formation and development of both the Saratov region innovation system and the national one.

It should be noted that all documents described above were developed in the Saratov region *before* the passing of "Principles of Russian Federation Policy in the field of the national innovation system (NIS) development up to 2010". This is evidence of the outstripping development of national innovation system elements on the regional level (bottom-up development).

NIS infrastructure development in Saratov region

The regulation and management of innovation activity are not restricted only by the adoption of legislative acts. They demand an integrated system of corresponding measures and structures. The development of innovative activity demands the formation and improvement of adequate infrastructure. In Russia this process is carried out by the creation of science parks at universities.

"Volga-Technics" of the Saratov State Technical University (SSTU) is a science park comprising about 20 innovation firms, producing and commercializing new products and rendering more than 30 services in the certification and protection of intellectual property for industrial enterprises. In 2002 the SSTU science park "Volga-Technics" carried out R&D at a cost of more than 50 million roubles. At the same time an innovation infrastructure of the science park is developing. This system provides patenting, expertise, marketing, advertising, information and personnel services for innovation firms.

Our next step is to create Innovational–Technological Centres (ITC) and Innovational–Industrial Complexes (IIC). The aim is to coordinate all structures and innovation processes in the region. These units are organized jointly by one or more enterprises and research and educational organizations. Their aim is to carry out innovation projects and programmes. The development of these structures involves industry enterprises, especially during the first steps of the innovation process. The cooperation between universities and enterprises ensures not only the effective promotion of new R&D results, but also staff training.

In recent years, based on the cooperation of the Saratov State Technical University with industrial enterprises, some Innovational–Technological Centres (ITC) and Innovational–Technological Complexes (IIC) were introduced, which shows great efficiency in the development of new products and technologies: ITC "SBP Corporation – SSTU" (1999), ITC "SAP Corporation" – "SSTU" (1998), ITC "SRPE Contact" – "SSTU" (2001), SSTU ITC of Intellectual technologies (2001), ITC "Rosdortech" – "SSTU" (2002). The ITC of "Saratov Bearing Plant" Corporation with Saratov State Technical University (1999) and the ITC "Contact" of state research and production enterprise with Saratov State Technical University are both excellent examples of university and industrial enterprises cooperation.

State support of domestic innovation projects implementation in Saratov region

Corresponding to the priorities of the development of the region, districts, certain enterprises and industries, innovative programs and projects are being developed. Universities, being suppliers of intellectual products and qualified personnel play an important part in their realization. As an example of an effective regional innovative programme, in which Saratov State Technical University takes part, is the Governor Programme "Energy Saving in Saratov Region for 1998-2005". Earlier consolidated non-budget funds of energy supplying organizations were the first sources of financing the programme. Now the programme has shown itself to be effective and it is financed from the regional budget. In 2002, 20 million roubles were allocated from the regional budget, together with 80 million roubles of outside funds, to finance the programme. In 2003, 30 million roubles were given from the regional budget and 120 million roubles of outside funds. The future realization of this program is expected and after it is finished, the development of a new regional investment program entitled "Energetic efficiency in the Saratov region up to 2005-2010" is planned.

Financing innovative activity in the region is also carried out from regional and federal sources. At the federal level, funding is carried out through programmes of the Ministry of Education and the Ministry of Science, Industry and Technology as well as through different funds. In particular, the Federal Fund for Supporting SMEs in Science and Scientific Services finances more than 40 innovation projects costing more than 60 million roubles in 2003 in the Saratov region. Out of these, 12 projects at a total cost of 10 million roubles belong to researchers of the Saratov State Technical University (Fig. 3).

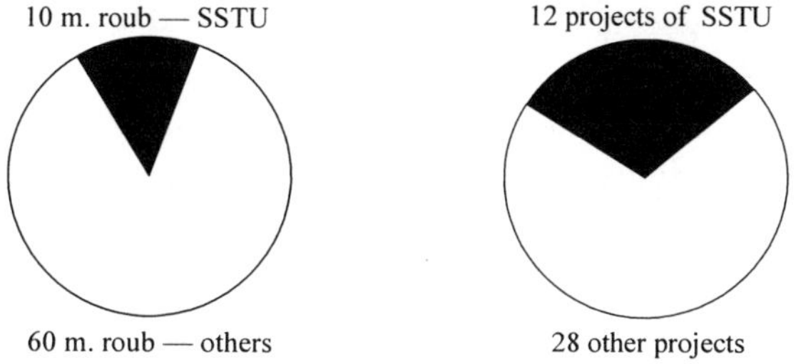

Fig. 3: Financing from the Federal Fund of Supporting SMEs in Science and Scientific Services within Saratov region (2003)

Recently the Federal Fund for Supporting SMEs in Science and Scientific Services announced the "Start" program, which will finance the start-up of new technologies. It can be seen as a programme for the creation of a market of proposals for venture capital. The main task of this programme is to attract venture capital into SMEs' innovation projects which will be supported by the Fund. A team applicant should be oriented to organize innovation business. Company grant-holders will continue to be self-managing, but monitoring and consulting are supposed to be regularly conducted. The Fund hopes to finance 400 projects, using about 3-4 millions of roubles in the framework of "Start" program. Saratov State Technical University (SSTU) plans to file some 10 applications under this programme.

As regards the regional budget, for example, in the 2003 Saratov region budget a separate category for financing innovation projects was included, to which 1 million roubles was allocated. Other regions also earmark funds for this purpose.

The government of the Saratov region has accepted the procedure of the formation of regional investment programmes. They are indicated as a separate line in the regional budget. In 2003, researchers from SSTU were carrying out 3 large-scale research projects in the field of ITT, electronics and energy-saving, with financing from the Saratov regional investment program. The total volume of financing for these projects from the regional budget was 710,000 roubles. In 2004, SSTU researchers have been carrying out a project on the development of new materials for health in the framework of the regional investment program.

Problems and barriers in innovation activity

Despite actively working on the Saratov region innovation system formation, there are big problems yet in this field. One of the main problems is the absence of effective mechanisms of transfer and commercialization of intellectual property, including that property which is created at the expense of the federal budget. The infrastructure of the transfer and commercialization of intellectual property is very weakly developed. At the present time there are only a few Russian consulting firms which specialize in this sphere, and few local networks. Most of these are in Moscow and St.Petersburg, but large parts of the Russian regions lie outside the innovation diffusion.

One of the ways of solving this problem is to create regional centres of intellectual property transfer in large industrial regions on the base of leading innovators including universities. In the future it will be possible to create a network of these centres similar to the Innovation Relay Centres Network in the EU. We would like to carry out this work in Saratov region in cooperation with EU partners.

Conclusion

In the 21st century innovations are one of the most powerful factors of social, and in particular, regional development. Their efficiency depends on the existence of adequate institutional and financial mechanisms, which are important elements of the innovation system. That is why the creation of such a system, on any level, requires their generation.

In this paper the processes of generation and development of institutional and financial mechanisms in the development for regional innovation system in the Saratov region have been described.

References

"О Совете при Президенте Российской Федерации по науке и высоким технологиям". Указ Президента Российской Федерации от 8.11.2001. № 1301.

Основы политики Российской Федерации в области развития науки и технологий на период до 2010 года и дальнейшую перспективу // Поиск. № 16 (674). 19 апреля 2002 г. С.8-10.

"Основы политики Российской Федерации в области развития национальной инновационной системы на период до 2010 года", февраль2004 г.

Закон Саратовской области "Об инновациях и инновационной деятельности" (новая редакция, принята Саратовской областной Думой в марте 2003 г.)

Стратегический план развития Саратовской области. – Саратов: 2000.

R&D Priorities in Innovation Policy and Financing in Former Socialist Countries
W. Leal Filho and P.S. Gramatikov (Eds.)
IOS Press, 2005

Chapter 11
Identification of S&T and Innovation Priorities and Real Policy of a State: the Case of Ukraine

OLEXANDR POPOVYCH, LIDIYA KAVUNENKO
Centre for Scientific and Technological Potential and Science History Studies
(STEPS Centre)
National Academy of Sciences
60 Shevchenko Blvd
01032 Kiev-32
Ukraine

Summary

This chapter describes trends related to science, technology and innovation, also with reference to policy-making in Ukraine.

Introduction

One of the key components of science and technology policy of any state is the identification of science and technology (S&T) priority areas. The Law of Ukraine "On foundation state policy in the sphere of science and technology activities" [1] was passed after disintegration of the Soviet Union in 1991. According to this Law, the Ukrainian Parliament approves priorities of S&T development. These S&T priorities, adopted in 1992, cover the most vital areas of S&T development the significance of which is acknowledged worldwide.

It should be noted that in Ukraine almost all knowledge production spheres were attributed to S&T priority areas (Table 1). Such a wide list of S&T priorities was designed to protect the science and technology potential of Ukraine in conditions of severe economic crisis.

Table 1: S&T Priority Ares in Ukraine

№	New priority areas (as of 2001)	Priority areas adopted in 1992
1.	Basic research on most important problems in natural, social sciences and humanities	(no analogue)
2.	Problems related to demographic policies, development of human potential and building up civil society	Scientific problems related to building up the statehood
3.	Preservation of the environment and sustainable development	Environmental protection
4.	Most advanced biotechnologies; diagnostics and methods for treatment of most widespread decease	Human health
5.	New computerised tools and technologies for informatisation of society	Advanced information technologies, complex automation devices, communication systems

№	New priority areas (as of 2001)	Priority areas adopted in 1992
6.	Most advanced technologies and resource saving technologies in energy production, industry and agroindustrial complex	• Ecologically clean energy production and resource saving technologies • Production, processing and storage of agricultural products
7.	New matters and materials	New matters and materials

At the same time in accordance with Ukrainian legislation [2, 3, 4], state S&T programmes developed on a competitive basis and approved by the Government are the main tools for the realisation of S&T priorities. But the government, at the suggestion of Ministries, has formed the state contracts for the most needed developments by priorities also. However, our attempts to scrutinise the mechanism of realisation of S&T priorities and to reveal the relative advantages for the research and development projects (R&D) enlisted as priority ones led to dejecting results. The share of S&T programmes in the state (government) financing of science was 1.7% in 1993, in 1995 reached 7.3%, but since the beginning of 1996 it featured steady decline, in 1998 was 3.9%, in 2001 – 0.37% and in 2002 – 0.07% [5]. We would like to note that the state budget was not completely realized during the last ten years and the percentage of actual expenditures on S&T priorities in the total (government) financing is lower than on other ones. The distribution of the share of S&T programmes and state contracts on the S&T priority in Ukraine in the total government budget expenditures for science is shown in fig. 1.

This situation does not seem to feature rather an essential shift when the course on the innovation way of economic development was proclaimed by the President of Ukraine in 2000 [6].

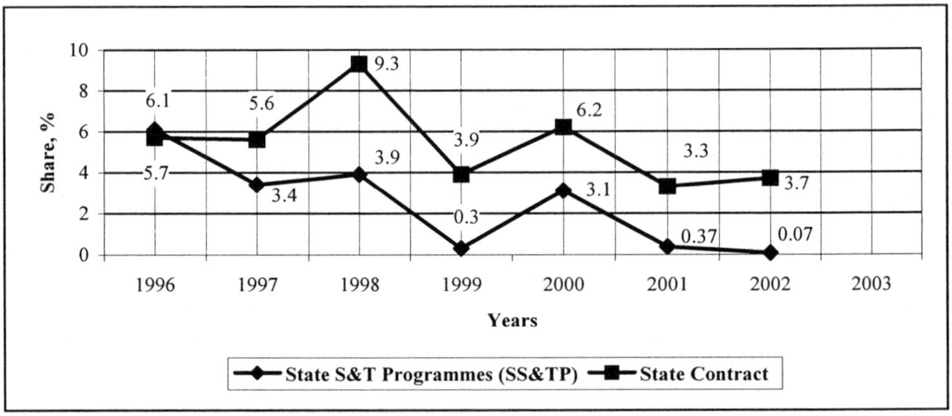

Fig. 1: The Share of S&T Programmes and State Contracts in the S&T Priority in Ukraine in the Total Government Budget Expenditures for Science

That is, we have to admit that the S&T policy of the state is implemented by overtly disregarding the truth formulated by Gennady M. Dobrov, the founder of the Ukrainian science of science: "Experience has led to a deeper realization of the fact that the primary essence of the category 'priority' combines 'leadership in achievement' and 'preference in logistics'. To rely upon leadership without the assurance of preferences means utopia, while preferences without accountabilities to assure this leadership mean privileges corrupting the healthy organism of the implicitly competitive development of science" [7].

Reasons for this phenomenon are various, such as:

- disregard of the complexity of mechanisms for the realisation of S&T policy, coupled with the crisis of public administration [8];
- the proclivity of the national bureaucracy to imitation of vigorous effort instead of searching for concrete measures for the consistent implementing of any policy [9];
- a number of objective factors relating to poor grasp of the notion "S&T priority" and its functions within the mechanism of S&T policy.

At bottom, interpretation of S&T priority is radically different in Ukraine. The specialists, who proposed these priorities, supposed that S&T priority areas would be supported by government. Lawmakers who adopted S&T priorities thought that S&T priority areas would receive special assistance, and representatives from executive power who funded S&T priorities awaited the special instructions.

As a way out, a methodology of structuring and hierarchy of the priorities was proposed, based on the mechanism of realisation and the relevant level of support to the S&T priorities.

A logical scheme for such structuring is shown in Figure 2. It means that before identification of state priority it is necessary to have a clear goal for what exactly is to be done and what is supposed to be done with it on every level. So there must be definition of the relevance between goal setting and goal achievement. Then it is also necessary to reach agreement upon it on all administrative levels. This is because a type of the priority is automatically linked with its implementing mechanism. A priority, if defined in a wide way, cannot be supported by a powerful stock of resources, as the resources would not suffice even with a well-off state.

However, such S&T priorities are valid and they are de facto required. Because, in addition to the meaningful question: "*priority relative to what*?", which urges one to think that the very notion of a priority implies the availability of less priority areas of research. But impossibility to concentrate large resources in a wide priority area does not deny its moderate but nevertheless essential support, e.g. assurance of stability in the basic funding of relevant institutions, tax and credit privileges, moral support etc.

Given such a level of priority, we believe it is acceptable approach when support to basic (fundamental) research *per se* is affirmed as a state priority, as practised in Ukraine in 2001 (see Table 1), even if it implies that a state with rather a moderate stock of S&T resources is able to assure the advanced performance of only a part of the vast spectrum of modern basic research. Hence in practice such a priority is realization in the form of support to research teams and academic schools that have won authority and international recognition. Besides that, to affirm such an area as a priority one would require a rather powerful state fund for basic research that would be in a position to support the development of research defined as new on the basis of expert opinions.

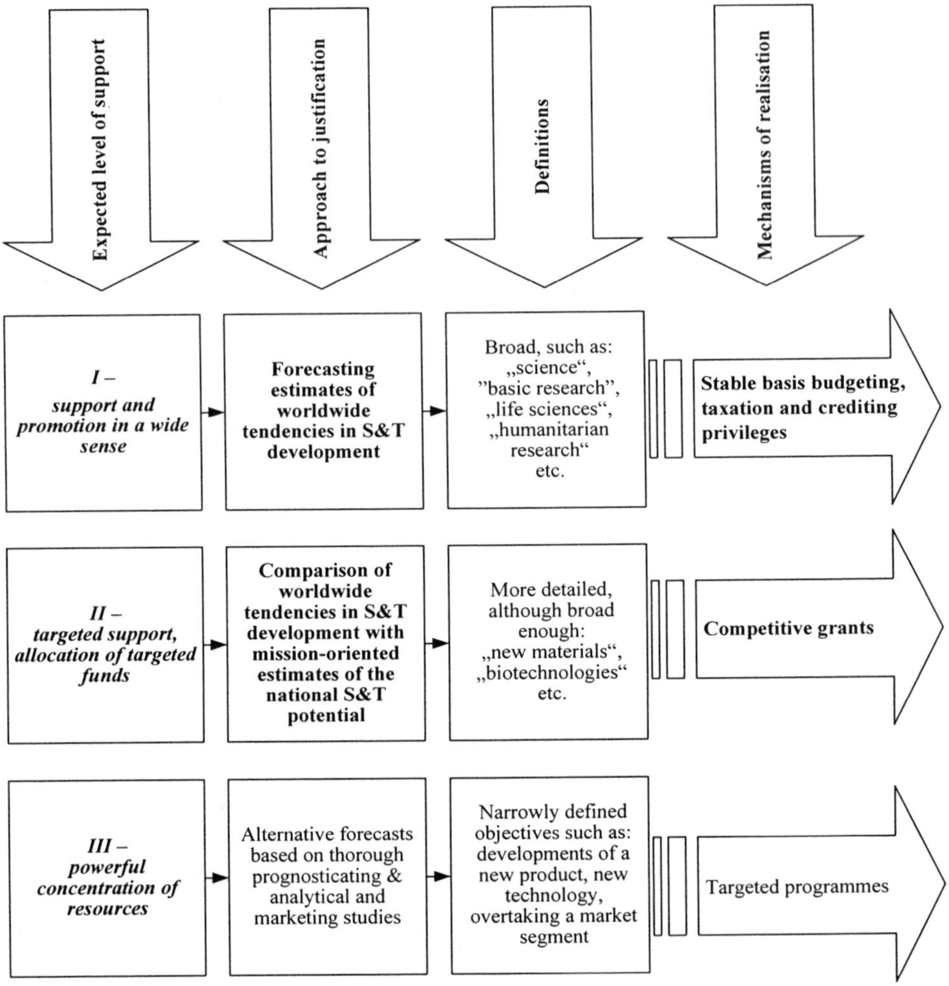

Fig. 2: Logical Scheme for the Formulation of a Hierarchy of S&T Priorities

The second level of priority is dependent on actual possibilities of the targeted allocation of available additional resources. Priority areas of this type must be identified on the basis of prognosticating and analytical studies of scientific and technological developments and their comparison with mission-oriented estimates of the S&T potential of the country. We believe that the principal mechanism to implement this type of priority must be competitive grants or state contract – for R&D representing a specific interest of the state. We believe that widespread practice to refer support of so wide a spectrum of R&D and realisation S&T programmes is erroneous. This is because while these S&T programmes are being referred to as targeted, they nevertheless more often than not cannot be structured as individual objectives and tasks.

Targeted programmes are required logically for the third level of priorities that presupposes powerful concentration of resources on the breakthrough R&D areas in world markets for high-tech and R&D-intensive products. To identify such areas, an S&T forecast or a mission-oriented study of the available intrinsic S&T potential must be supplemented by marketing studies and economic calculations of the performance of the S&T programmes.

Given these conditions, these S&T programmes can be deliberately targeted, and as they are being implemented, advantages of the mission-oriented arrangement may fully show.

It seems to us almost actually (not for Ukraine only) like a return to the original understanding of a programme-goal approach, as an ideology of advisability and as a method which offers an optimal way to the goal in the process of realisation of the S&T programmes.

If one goes back to S&T priorities adopted by the Parliament of Ukraine and shown in Table 1, it should be noted that it is the methodological confusion in defining the level of priorities and their implementing mechanisms that seems to us to be one of the principal reasons behind their extremely poor performance. As can be seen from Table 1, they all are wide enough and refer to either the first or, at best, to the second level. However, according to the law decrees to use state S&T programmes as their main mechanism of realisation of S&T priorities, e.g. the one that applies only to the third level of priorities, which is narrower and more focused. It is clear that the development of an effective S&T programme on any of the priorities was logically impossible. As for any clear objectives and results that could draw the attention of powers, these were also impossible to formulate in such blurred and so poorly funded S&T programmes. Due to this, the very notion of "priority area" eventually lost its primary sense and merit.

Note that in addition to the proposed one, there can be somewhat different (although, probably, analogous in many ways) logic for drawing up a hierarchy of S&T priorities, based on the division of tasks for long-term strategic priorities and those which solutions are expected in medium-term and short-term future. It is clear that strategic tasks allow for a broader definition corresponding with the first level of hierarchy given in Figure 2. Medium-term tasks, accordingly, correspond to the next levels.

Given this approach, we proposed to design a system of priority areas of innovation activities in Ukraine, adopted by the Parliament of Ukraine as a law [10], which may be seen as a move forward in realising the priority-related approach to policy-making and assurance of more systemic approach in its implementation. Thus, this law envisages the design of a hierarchy of innovation priorities consisting of the four levels.

Table 2: Hierarchy of priority areas for innovation activities in Ukraine

№	Hierarchy of priority areas for innovation activities in Ukraine	Duration (years)
1.	Strategic priority areas for innovation activities in Ukraine	10
2.	Medium-term priority areas for innovation activities in Ukraine	3-5
3.	Medium-term priority areas at branch level	3-5
4.	Medium-term priority areas at regional level	3-5

By this law, the proposals on inclusion of an area in the priority nomenclature are compiled on the basis of prognosticating and analytical studies. However, because as of the date of passing the law no study was available, and if priorities are not adopted, their implementing mechanism envisaged by the law on innovation activities cannot be launched, the same law [10] decreed the strategic areas for innovation activities over 2003-2013 given in Table 3. We cannot say that we are fully content with strategic priority areas for innovation activities, as they are in a way incidental. This is because, contrary to the legal procedure, they did not come from any thorough prognosticating and analytical study but were compiled by a rather narrow group of specialists. Nevertheless, at the stage of building up the innovation system and its mechanism of realisation of the innovations policy in Ukraine, we believe it may play a positive role. We are especially pleased to say that following our proposal, fostering of innovation culture in society was legally defined as a national strategic priority of innovation activities (e.g. at the level of the state). The innovation culture is defined in this law as "a component of the innovation potential featuring the level of educational, cultural, social and psychological preparedness of a person and the society as a whole to perceive and creatively implement ideas of economic development on innovative basics" ([10] – Article 2).

Table 3: Strategic Priority Areas for Innovation Activities in Ukraine

№	Strategic priority areas for innovation activities in Ukraine
1.	Modernisation of power plants; new and renewable energy sources; advanced energy saving technologies
2.	Machine-building and device manufacturing as the basis for technological renovation of all the manufacturing industries; development of high quality metallurgy
3.	Nanotechnologies, microelectronics, information technologies, telecommunications
4.	Improvement of chemical technologies, new materials, development of biotechnologies
5.	High-tech development of agriculture and processing industry
6.	Transport systems: construction and reconstruction
7.	Protection and improvement of human and environmental health
8.	Fostering of innovation culture in society

On each of the aforementioned strategic areas the law specifies medium-term S&T programmes at the state level, and envisages drawing up sector-based and regional ones. These innovation priorities may be criticised for inconsistency and incoherence of the areas referred to, and for their overlap with S&T priorities. But it is certainly a move forward – towards a systemic vision of policy-making in this sphere.

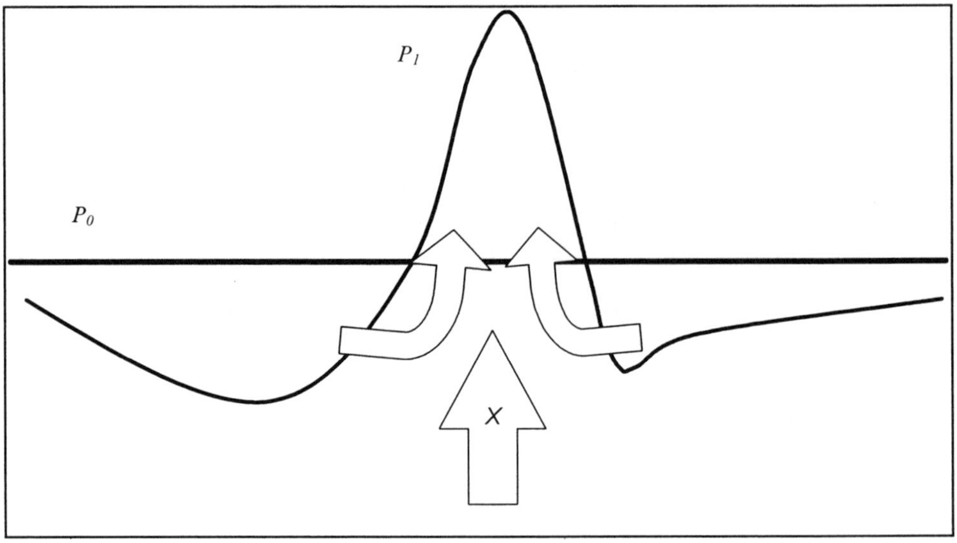

Fig. 3: Illustration of Regrouping of the Scientific Potential on a Priority Area (P_0 – initial level, P_1 – after re-grouping)

From the systemic point of view, it is very significant that the priorities are identified at different levels both by the range of the problems involved and by their implementing mechanisms. At the same time it should be emphasised that, playing a specific role in the overall hierarchy, priorities at different level also interact and have causality links.

One should not ignore the fact that building up the S&T potential and its concentration on the selected area is *per se* a rather inertial process requiring quite a long time. And while in the past it was not a limiting factor, as time lag from a scientific discovery to practical implementation used to be large, in our dynamic time this has already become a serious problem.

So virtually all the known cases of the rapid concentration of the scientific potential on a selected area (an example is the nuclear problem) were in fact confined to involving specialists from related fields. This may be illustrated schematically, as shown in Figure 3.

If an initial level of the scientific potential is denoted as $\boldsymbol{P_0}$, then concentration of the effort on an area X would lead to its new distribution $\boldsymbol{P_1}$ that can occur in no other way, at least in the nearest future, except by attracting resources from related fields.

And this means that the availability of such areas having either low insignificant priority or being related with broader priorities represents a necessary condition for the formulation of a priority.

References

1. Law of Ukraine “On foundation state policy in the sphere of science and technology activities” from 13 December 1991. In: Information of Verkhovna Rada of Ukraine, K.: 1997, P. 16–30.
2. Law of Ukraine “On Priority Areas of S&T Development” from 11 July 2001, N 2623-III. In: Information of Verkhovna Rada of Ukraine, 2001, N 48, P. 253.
3. B.A. Malitsky et al. Vital Issues of Methodology and Practice of S&T Policy. Kiev, 2001.
4. Law of Ukraine “On Scientific and S&T Activities” from 1 January 1998. In: Information of Verkhovna Rada of Ukraine, 1999, N 2-3.
5. A.S. Popovich. The Place of Priorities in Implementing the State S&T Policy of Ukraine. In: Science and Science of Science. 2001, N 2, P.P. 65-73.
6. Ukraine: Moving Forward into the 21st Century. Strategy for Economic and Social Development over 2000-2004 (2001 Message of the President of Ukraine to the Verkhovna Rada of Ukraine). In: Governmental Courier. 23 January, 2000, N 34.
7. G.M. Dobrov. Science of Science Kiev, Naukova Dumka, 1989, p.277
8. A.S. Popovich. Mechanism for Implementation of S&T Policy as a System. In: Science and Science of Science. 2002, N 3, P.P. 36-46.
9. A.S. Popovich. Bureacratisation of Society and Scientific Life as a Factor Hampering Innovation Development. In: Science and Science of Science. 2003, N 4(20), P.P. 79-96.
10. Law of Ukraine “On Priority Areas of S&T Development” from 16 January 2003, N 343-IV. In: Information of Verkhovna Rada of Ukraine. 2003, N 13, P.P 354-358.

R&D Priorities in Innovation Policy and Financing in Former Socialist Countries
W. Leal Filho and P.S. Gramatikov (Eds.)
IOS Press, 2005

Chapter 12
Innovative Territorial Security Policy

IVAN VELITCHKOV
Southwest University "Neofit Rilski"
1, Georgi Izmirliev – Makedoncheto Sq.
BG-2700 Blagoevgrad
Bulgaria

Summary

This article discusses some current tendencies in the field of territorial security which require a lot of innovations, a serious investment in order to get successfully under control the manifestations of terrorism, hackers, and the violations of the social order, as well as to counterstrike the wrecks, disasters and catastrophes.

Introduction

Together with national security, territorial security attains increasing significance. Security in the administrative-territorial units (mayoralty, municipalities, and communities) becomes more and more important after the accumulation of many people in the administration centres.

When research and developmental priorities of innovative policy as well as their financial mechanism are being discussed at a such level and with the most eminent specialists, "territorial security" as a prior problem should be carefully put up, examined and estimated. It just requires urgent principled innovative action along with serious investments. A very short time ago the term "security" was widely used at micro level with the qualitative definition "national". After ending and overcoming the "Cold War" and the unification of separate states from various regions on different bases, it is understandable that higher attention is required to follow, estimate and positively alter "territorial, regional security". More and more often events, catastrophes, natural disasters, failures, acts of terrorism take place in a particular territory in a particular large city or settlement centre in a particular direction and local expression, without threatening national security (New York, Madrid, Kabala, Balkan and Near Eastern states and many others).

Innovations and renovations in their essence, yet with different terms, have always been a motor of human progress and have been setting a different pace of social prosperity. Their dynamics, importance, speed, comprehensiveness are now, at the beginning of the 21st century, becoming increasing and incredible, and acquiring fatal significance. Since millennia it has been clear to us that "panta rhei – everything is in a state of flux and motion", meaning "Everything changes" and it is changes that determine our present discussion. There are numerous reasons and motivations that have united the leading specialists around the "round table" to discuss the research and developmental priorities of innovative policy on NATO's initiative. Without underestimating the (always emphasized in innovative policy) leading positions of economical, social, ecological interests, aims, effects, we should put along with them those of "local, territorial security", and the financial mechanism as well, considering the wretched economical status of our country and some other ex-socialist countries too.

Challenges of the new century

At the beginning of the new century and the new millennium, the world and any state, region, territorial unit, any company, any person cannot afford not to pay attention to the increasing insecurity and take good care in physical defence matters, informational protection against escalating terrorism, crimes, catastrophes, and surprises. The separate individual and the whole of mankind are facing the serious question of survival in present and future conditions. Over six billion people live and work on earth. They breed offspring, create values, produce material and spiritual goods, and lead their own way of life. In this elaborate and crucial process, two factors are becoming increasingly significant and require special attention:

- Increasing density of population and concentration of large human masses in big cities – urbanization.
- The rapid qualitative changes in the human race of today, associated with their intellectual growth informational reload, armament with up-to-date technology in their living standards, leisure time and working place.

Undoubtedly, all this demands new dynamic and constantly changing lifestyle subordinate to striving for survival and material welfare. Elaborate and dynamic processes are running in every administrative-territorial unit – lots of material valuables remain inactive, unemployment is growing and so is crime. The number of incidents disturbing people and local authorities is rising. This demands that the local administration cooperate with the police in order to maintain the security, public order and peace of its citizens.

The running social-political and social-economical changes also demand reorganization of every activity within the state in order to maintain balance in social life.

Obedience and observation of the basic principles of legitimacy and legal order regulated in various statutory and sub-statutory enactments (decrees, instructions, resolutions), as well as respect for the rights and interests of the citizens of the Republic of Bulgaria. These are the main obligation of every Bulgarian citizen and are regulated in the Constitution – the basic and superior enactment, valid for the territory of the Republic of Bulgaria.

That is indeed, what requires more thorough knowledge, interpretation and understanding of these enactments of the organs of local authorities and organs of the Ministry of Internal Affairs, so that law-breakers can be sanctioned.

This concerns social benefits such as internal order, security and peace in the state and in every administrative territorial unit, for which public expenses are made. Based on public will, in other words on citizens' or civil structures' preferences for the consumption of certain public goods and services, public authority and its administration determine their type and quantity, which are offered and conceded to citizens and civil structures or to society in general. The issue "order and security" is put down along with "unemployment" as a priority before central and local governmental organs. Public authority and public administration are relied on for urgent alterations in this insecure and fearsome direction.

In the process of the democratisation of society there is a growing role of acts which are extremely dangerous and lead to the brutal violation of social order, finding expression in frank disrespect for society, disturbance of the peace, rest and work of the country's citizens.

Division of power and slow-going reforms of the legislative system in the process of democratisation have led to an increase incrime in some regions, communities and particular settlements. And besides violent acts, intellectual crime makes an appearance.

The need of making clear the common concepts and formulations, connected with security via philosophical, psychological, economical, juridical and cybernetic interpretation, is dictated by the new realities. That is especially valid after September 11 2001. Mankind condemned terrorism and led an open war against it.

The former ideological enemies turned their backs on their contrarieties. On 28 May 2002 Russia and NATO signed a contract in Rome. Thus formally the fifty-year Cold War actually ended.

The irreversible way towards a market economy began generally to change the attitude of people in our country too concerning ownership, production instruments, distribution and redistribution of goods. The process of privatisation caused a radical turn of former defence and self-defence systems. The focus was moved from state to private security agencies. The word "security" itself, often used demagogically by many politicians, has turned into abstraction for ordinary people and plunged people to seek salvation in religion, mysticism and superstition.

Social-political and economical reality here at home and around the world gave rise to a new method of approach in research considering the survival of the separate individual, on one hand, and the whole of humanity on the other. The USA and Japan have officially declared the beginning of a new era of "cybernetic" security.

Reconsidering the former theory and practice, associated with security, the conclusion to make is that it is necessary to set up a new science relating to the survival of humanity. This science carries the common concept of "security". It also includes many different forms (individual, family, and group, corporative, territorial, national, continental, and global).

The thorough examination and rationalization of every one of those forms separately will undoubtedly help the law-giving system of all states with the juridical regulation of various different defensive and self-defence mechanisms. Such mechanisms would give people a chance to survive and prosper. After their birth, education and upbringing every human being who has come of age, gets into complicated relations and interactions with other individuals and institutions. Some of them he uses to create a family community and offspring and others to create group or labour – juridical interrelations and third – international contacts, agreements etc. Under certain circumstances man appears in his capacity as a physical person, and under different circumstances as a juridical person. He could live for some time in some social-economical conditions and afterwards in others. Sometimes he is an object of various threats, but he could become a threat or menace himself. He could build and create but also steal, kill and destroy.

His family, ethnical and religious traditions and interpretation of moral, spiritual values and security are also important.

All these characteristics of activities of individuals and their behaviour should be reported and be an object of special research. The cornerstone, however, in which a lot of the prominent minds dealing with security issues trip, are the unique abilities of man to get, acquire, generate, preserve, use, carry, sell or destroy valuable information.

It is the great variety of dynamically changing forms and status (not only the one of the individual, but of the existing civilizations as well) that makes it difficult to formulate a universal definition, that would envelop the essence of security in its entirety.

Security of Innovation

Scientific literature does not have a precise definition of security so far. Usually security is identified with similar words (safety, stability, self-defence, defence, protection etc.) related to the separate individual, system or state.

The most common meaning of the term "security" could be: *Security is the realized, rationalized, purposeful, functional and dynamic defensive and self-defensive system of survival and stable existence of physical and juridical persons, in the duration of a certain period of time and under certain social-economic circumstances.*

Therefore security in its most general (global) sense is a science for the survival and prosperity of physical and juridical characters as well as of humanity as a whole and as a community. Despite of this, security is an ideology and high art, "operative master ship", which must be studied not only by those who have devoted themselves to this profession, but also by people of any age. Security as a science for the survival of individuals and the humanity as a whole is not only possible but even necessary to include all other sciences, especially the nature-researching ones and to report all scientific-technical, technological and so on innovating decisions and achievements. This is also obligated by the fact that people are constantly in contact and cooperate (positively and negatively) not only with each other but also with the surrounding world – ecological environment, space, and universe. And this world is not at all well intentioned when a negative influence is caused by the humans over it, especially when a war or an eventual use of atomic, chemical, bacteriological etc. weapons is expected.

There is also a quantity of fatal natural threats to people – volcanoes, earthquakes, inundations, typhoons, hurricanes, droughts etc. A crash between the earth and comets, meteors and meteorites is also possible.

In such situations a unification of the efforts of the entire mankind is compulsory for effective counteraction to them. This can happen only on the basis of a thorough and reliable security system. And this is even more urgent now when the new face of "the world's evil" is a hacker, a terrorist or an Islamic fanatic who threatens territorial security.

In comparison with the other types of security, the territorial one has a special place, role and specification. It represents a complicated conglomerate of state, territorial and private interests. It is characterised with the leading role of the political power, which has won the elections in a particular administrative-territorial unit. It carries the signs of the individual, the family, the group and the firm security. Because of this reason the territorial defensive and self-defensive systems that have already been built and are functional are to some extent a micro-model of national security and an essential part of it.

Since the individual, the family and group security have for their basis criteria for diversifying the human factors in firm and territorial security, a priority is given to the institutions and their executive organs, which gives a guarantee of security and order in firms and administrative-territorial units.

If an administrative-territorial unit – a region, community or a town council is considered a complicated and distinguished system, it cannot think of the "security" problem and territorial security with negligence and disrespect. We cannot consider it only a fashionable, temporary or unrealised problem. This is an essential and crucially significant matter for every cybernetic system.

Territorial security can be defined as an organised objective and lawful activity of the local governing structures and authorities, which in cooperation with The Ministry of Internal Affairs organs, all public institutions and also the citizens ensure security, reposefulness and order on its territory, preserve the lives and the properties of people and exclude possible harms caused by crimes, disasters, damage and catastrophe.

Territorial security has never missed its popularity and significance, but now in the new social and economic conditions it gets a priority in society, attracting the interest and the attention of governments and all kinds of authorities.

People with their human communities are born, function and die in a distinct territory. It is a natural base for their lives, activities and their death. From this outer realistic fact follows the human dependency on the concrete place where humanity interacts with Nature and with other individuals. That is why the territory is at the basis of the mere structuring and organisation of public life. That is why the territorial organisation of institutions is approved by history as the system-developing, as the organising influence of a given natural perimeter over people's lives and the human community they create.

Exactly those borders, the definiteness of the country's territory and of the administrative-territorial units in it becomes a factor that limits the relatively constant structure of the population, and from this point of view a factor for a higher, group and administrative organisation of human activity. At the same time the constant territory gives a possibility of the central authority with the help of its local institutions to control the deeds and the behaviour of the population, as well as to regulate the relationships, order and security directions and dimensions in the desired direction.

The territorial organisation of a country has a huge significance for the right organisation of the political, economic, social and cultural life of society. That is why it has to be built on a scientific basis and in full consideration with the interests and the will of the local population, which is an enquiry in our Constitution (Article 36, Paragraph 2). The administrative-territorial structure suggests also the subdivision of the country's territory into units with defined legal status.

The community is a form of social organisation, which comes from the vital necessities of people. It is an administrative-territorial unit, which includes in its boundaries land of one or more settlements. This is the territory which is inhabited by the population of the community, the territory on which are situated most of their properties. The community is first of all a consulate of people who inhabit one certain territory and who are gathered to govern, defend themselves and defend their common interests. The territory itself is important, but people in it are more significant, which defines the territorial commonwealth.

Communities are local institutions and their organs decide questions from the point of view of local significance. In spite of this they execute important tasks of different character such as social, agricultural, cultural, medical, educational and structural planning, tasks concerning order, security, defence and many others. The most dynamic and defining factor in every community is its population. These are the citizen's – Bulgarian and foreign people that live on this territory. The juridical connection between the community and the separate individual is the institution of right of residence.

Any Bulgarian citizen is obliged to settle himself or herself in one defined settlement, where his or her constant address registration has to be. Right of residence entails obligations. A consequence of the movement of people is the increasing violation of the social order, the frequent hooligan acts and the acts of terrorism, and burglaries and damage to personal, municipal and state properties.

The preservation of the social order is a basic function of the state and its administrative-territorial units. This creates the conditions for security and freedom, for protection of its legitimate interests. The municipality forms and conducts common policy for the protection of the social order and security and on its territory. In this connection it coordinates its activity with the police, the court and control agencies on its territory, which is a precondition for higher effectiveness in the struggle with criminality and the other transgressions of the law. Its duty is to provide protection for the communal funds and to specify their usage, to ensure calmness in the lives of its citizens. This is achieved with regulations, which the town council accepts, as well as through the acts and actions of the municipal police.

The protective functions of the municipalities include the preservation of agrarian properties, effected with the help of the field guard and with other means. Great are the concerns and for the preservation of municipal and state property from fire and damages, protection of the population and territory from disasters, damages and catastrophes.

The municipality takes special care to prevent crimes committed by children and young persons, takes responsibility for social work with children who are prone to criminal behaviour, to drug addiction, to prostitution and other immoral acts.

All this gives reason to speak for the synonymy of territorial and municipal security, as well as the specificity and mechanisms for achieving it.

As for the greater part of people's everyday activities – bank operations, travel, post services, shopping, entertainment and others – they can be registered with the help of different information technologies, among them computer technologies, and it becomes technically easy for them to be catalogued and observed. The increase of the scale of local and global computer networks increases the possibility of the abuse of information by individuals, and of wrong actions from the administration.

In accordance with the "Strategy for development of a modern administrative system in Bulgaria" adopted by the Council of Ministers, certain efforts for optimising these methods have been made, and the state and local administrations serve the country's citizens and the national business with reference to them. These are directed to:

- establishing European norms for making information accessible to the public, while the security of the data and the main human rights are guaranteed;
- renewal of the functions, the structure, the products and the utility services of the administration according to the new information and communication systems and formation of favourable conditions for a stable development of society;
- establishment of mechanisms for guaranteeing the citizen's communications rights – a right to the free circulation of opinions, a right to information and the security of personal data.

The computer services offered by the different sections of the public administration should be approachable, easy to use and cheap for the taxpayer. The realisation of the National Strategy for transition to an informational society should lead to a harmonizing of the relations between the administration, the population and business, as well as to the growth of the democratic control of the population over the country's government.

All this, together with the state of the society in the future will make it possible to speak of a cybernetic society. Society is a complex and extremely dynamic system which makes progress with the help of the newest modern technology, with powerful computers and control system centres which allows the better and more precise use of the principles and rules of the cybernetics – hierarchy, rationalism, formalisation, effective regulation activity.

A reason for this is both the progressive tendencies and needs, and the increased and constantly fed cyber-fears.

It is not by an accident, that the new face of "the world evil" is a hacker, a terrorist and an Islamic fanatic. The attacks against the Internet infrastructure also show the prospects of a total collapse of the global net. The DOS attack on the Internet which caused the failure of eight of the total thirteen servers in 2002 is considered to have been the most complex and serious threat in history, and made obligatory the search of a new security scheme and system against the cruel cyber terror.

This is even more necessary now in a territorial (municipal) aspect if we have in mind what rich, working and yet modern information technologies, which is the municipalities and the districts, and how much personal information equipment is worked with and will be worked with. Fears for security and personal inviolability in the network are not groundless and at least for now it is in hands of the individual user. It depends on the people to ensure that they are reliably protected, because it is not enough to believe that the problem is not down to the separate individual. It cannot and it must not be forgotten that information is power, power is money and those who possess the information in the social or information society, possesses valuable commodities – the possibility of trading with full knowledge, and more actively in territorial and regional aspects.

Attention to the economical situation in the country is of big importance for the finding of a maximally precise formulation of the concept "defence" (security) on the conceptual level and as a system of purposeful actions (defence measures) in our specific condition. The question of the protection of the population is directly linked with survival issues in conditions of critical situations, with disasters, damages, catastrophes and others. The concept defence is key regarding the system of measures, which the state authorities and organization needs to undertake so as to guarantee the security of the population and prevent and liquidate the consequences of disasters, damage and catastrophes. In this linkage particularly important is the question of financing, and well provided for the activity of the defence.

With a decree of the Ministry council dated January 23 1998 a measure was adopted on the statute book for the organization of activities to prevent and liquidate the consequences of disasters, damage and catastrophes. In accordance with the new military doctrine and with the decree of the Ministry council no. 53 of March 2 2001, the "Civil defence" bureau has been transformed into the "Civil defence" state agency. This has brought about some positive changes in the active structure of the system of defence.

The organization, management, coordination and control of activities directed to prevention, reduction and liquidating of the consequences arising from disasters, damage and catastrophes as realized by the Ministry council, from the ministries, departments and agencies of local self-government and local administration, from trade associations and privately-owned companies, according to their competence.

The general management of the activity of defence is realized by the Ministry council. For the realization of defence activities by the population in critical conditions the Ministry council has set up constant commissions for defence against disasters, damage and catastrophes.

Financing of the innovation security system at local level

For the realization of preventive activity, mayors of the municipalities plans fund in their annual budget for fulfilling the envisaged measures of defence.

On demand of the chairman of constant commissions, the managers of the associations, organizations and competent institutions submit information, linked with the risk factors and results of the preventive activity. Normative basis for the conduct of preventive activity is laid down in chapter four of the statute book for organization and activity.

The decreasing of the dangers connected with the appearance of moments, concerned with the human factor depends on the effectiveness of the conducted preventive activity.

Moreover, insufficient financing makes the functioning of this scheme difficult, even impossible. That is why financing in the municipality is a guarantee for success not only in the preventive activity, but also in the whole system for defending of population in cases of calamities, damage and catastrophes.

Connected with this, effective financing is a guarantee for success of the preventive activity if it is to reach its main aim – defence and the mitigation of harmful consequences.

The activities connected with the defending of the population and the national economy in calamities, damages and catastrophes are financed accordingly from the state or municipal budgets, from the incomes of sale activity and presented charity and subsidy. The constant committees create a reserve of material means, needed fulfilment of the tasks for defending of the population and the national economy in cases of calamities, damage and catastrophes.

The municipal budget is the basis for conducting preventive activity. The municipalities in the face of the permanent municipal committees are responsible for planning the funds in their annual budgets for execution of events connected with the defence and organising and conducting of preventive activity for submitting and mitigating the harmful consequences of calamities, damage and catastrophes.

The financing of the preventive activity can be made not only from the municipal budget, but also from headed financial funds from the republic budget, following the resolution of the permanent committee for defending the population from calamities, damage and catastrophes.

There exist opportunities for the financing of projects and programmes for preventive activity with the help of the national fund for environmental preservation, administered by the ministry of environment and waters and science investigation funds.

The Ministries, the other departments and economy organizations submit funds to their factories from the Economy Risk fund, as well as free funds from other financial organizations, for redress of the damage and compensation for losses in case the factories do not have the last ones.

Funds for the opportune liquidation of the consequences of natural disasters and big production failures, as well as the restoration, strengthening and other activities, are provided annually in the republic's budget. The size is defined every year and depends on the amount of incomes and costs in the budget and the economical status of the country.

The more funds are planned and provided in the state budget, the more successfully will the activities for defence be conducted and the more effective the preventive activity in administratively territorial units will be.

A threat turns to a disaster when it comes over the human community and ruins its normal functions, when it affects people and provokes economic damage. There are calamities, disasters and catastrophes in all countries – poor and rich. But the ones that are worst harmed are the poor, mostly because of the absence of financial funds for timely preventive activity.

The Ministry of regional development and public works and the Calamities and Disasters fund can help stricken families with material means and with financial aid to citizens affected by natural catastrophes. The State Savings Bank can give loans bearing no interest to people affected by natural catastrophes, from a limit of flat buildings for restoration and repair of buildings with a term of payment of fifteen years.

The other important source of financial funds for defence activity is the municipal budget. It is a part of the national finance system and has its own autonomy and independence.

The budget of every municipality consists of an income and a cost part. The income part of the budget is formed of sole incomes, state incomes remised to the municipality, drawn in funds, subsidies, subventions and other sources.

Sole income resources. They are the local taxes – tax on inheritance, tax on real estate property, tax on charity, vehicles and other. The local counters are with another sole income resource – in this category are charges for refuse collection, for use of markets, market places, nursery schools and others.

Other incomes are those from concessions. According to the concession contract, the government shifts the full responsibility for delivering infrastructural services in a particular territory to the private concessioner.

Subsidies and subventions from the republic's budget now determine the destiny of many municipalities. The subsidies are transfers from one public body to another, with submission of funds under certain conditions. The general subsidies are defined under objective criteria and are confirmed with a Law for the State Budget of the Republic of Bulgaria for the relevant year. The strategic subsidies are allotted for concrete cases for acquiring long-term material assets.

Another source of financial resources is in donations and sponsorship in favour of the municipality. They assure the accumulation of financial resources in the income part of the municipal budget. The collected funds are used for satisfaction of the communal-economical necessities and the realization of the function of the municipality, including the defence of the population from disasters and damage.

Municipal finances can form from private legal merits. The municipality as a juridical person can perform economic activities, and can cooperate with subjects under the private law, to obtain profit from economic turnover and in this way increase its property and acquire financial resources for its budget.

The municipal budget is a financial account for the profit and expense of the municipality for one budget year, which in terms of time coincides with the calendar year. In the expense part of its budget are foreseen the future budget credits for meeting the different activities of the municipality's administration, as well as for the maintenance of the municipality's agencies. The expenses of the municipality are made for coverage and satisfaction of the local necessities and problems, also including the execution of the measures envisaged for defence.

The municipality presented from the constant committee is obliged to allocate financial resources for defence measures, and conduct a preventive activity on the territory of the municipality.

The spending of the funds necessary for the realization of the measures for defence of the population, and the material values from catastrophes and damages and for conducting preventive activity, begins with the permission of the president of the constant municipal committee.

The expenses made from the municipality for rescue, restoration and repair work are reconstructed from purposeful financial resources planned in the republican budget.

In generalization it can be said, that the financing of the system for protection is directed towards enhancing the effectiveness of the struggle with major natural and technical disasters through prognosis, prevention, evaluation of the influence, development of earth surveillance technologies through satellites, monitoring of the environment and appliance of measures for decrease and avoidance of the consequences of disasters, failures and crashes.

The real model of the municipal organization for protection requires an extremely serious approach on the part of the entire managing and executive staff towards all these activities, including documentation of the protection system. All this is accompanied by a good financing of the municipal system for protection under present market conditions, but also to achieve its main aim – security and protection of the population, of the natural and territorial economy against various threats.

In connection with democratic changes in society and the even closer approach to the European structure, principles and lifestyles, as well as the realization of the main aim, which is set by our state policy for integration with the EU, there is a need for improvement of the organization of the work of the authorities, the police and the local authorities of the local self-governing for the safeguarding of the public peace.

In times of euro integration, democratisation and our opening towards Europe and the rest of the world a real balance sheet for our potentials and establishment of ethical principles should be drawn up. The answer should be precise because we must act against the negative incidents in our society. For a peaceful everyday life we need an active state government to lead a systematic struggle against all negative habits that contradict the law and the established values of the present.

It is imperative to lead a really determined and uncompromising struggle for protection of the public peace, of the prosperity, the peace, the honour and the life of the citizens. This is a struggle for a cheerful moral atmosphere and good organization; for peace and security that we so much need now, when there are such deep and qualitative changes in the society. This is a struggle for the individual and for the day after today. Because if we lack order, organization, discipline and supremacy of law, we will never be able to achieve the prosperity of society and the individual person, and we will not be able to achieve real democracy.

Conclusion

The peace, the property and the honour of the citizens of the Republic of Bulgaria are inviolable: they determine the directions of the perfection in the activities, the organization and structure of the police. This imposes an obligation of continuously analysis and perfecting of police activities, so as to respond better to the present changes on one side and the needs of people on the other. The national police is building now mainly on the guaranteeing of the life and security of the citizen, to maintain order, as well as to perform all its activities in such way as to hold onto the trust of the democratic society.

The guaranteeing of order and security presumes a serious innovative policy and activity, technical innovations, serious investments for informational preservation and for counter-striking the growing criminality, terrorism and violations of the social order.

References

Constitution of The Republic Bulgaria, State gazette (1991) No 56

Law for the local self-governing and the local administration

Law for the Ministry of Inner Affairs, State Gazette (1997) No 122

Law for the state budget for The Republic of Bulgaria

Marinovski, N. (2002) Police's actions in cases of calamities and wrecks, Sofia,

Security and social order – A programme of the Government of The Republic of Bulgaria

Velitchkov, I., Baldzieva, A., Gaidarov, I. (2002) Security, Sofia

Chapter 13
Economic Growth and Total Factor Productivity in Romania

ILIE SIMON
Academy of Economic Studies
Piata Romana, 6, Sector I
Bucharest
Romania

Summary

Based on the Solow model, this paper tries to analyse the different roles played by labour, capital and especially by what is called „residual" in the Romanian economy during the last 15 years. This last factor indicates a strong correlation between innovation and productivity and eventually economic growth.

1. Introduction

Looking at innovation from point of view of end results, such as economic growth, is a tempting approach (Romer, 1989). Economic growth in Romania, as in many other countries, was associated mainly with the process of industrialization, which has represented the bedrock of innovative process.

The attraction to industry has been very high, and not for a reason: in 1938 an employed person in Romanian industry produced a gross value 29.2 bigger than in agriculture. As a result of this orientation and efforts, during 1922 and 1938 the national income increased with an annual average rate of 4.3%, bringing the share of industry to 39% and of agriculture to 61% of total production. But then the war struck, and the national income in Romania in 1947 barely reached 35-40% of the level registered in 1938.

In order to analyze the contribution of innovation in Romania from the point of view of its contribution to economic growth after 1989, a short historical development will throw light on the present situation.

The result was that, after the efforts made during the first half of the last century, Romania was still occupying a bottom rung in the hierarchies of economic development on our continent.

During 1950 and 1989, Romania has continued the process of industrialization with a focus on investments, education and scientific and technical progress; industrial production increased during this period 48 times labour productivity in industry increased 12 times), contributing to a growth in national income by a factor of about 20 (Postolache, 1991).

The accumulation rate increased also gradually from an average rate of 17% in the 50s to a maximum of 35% during the five-year plan 1976-1980. As outlined by Dobrescu (1979), during the five-year plan 1971-1975 the accumulating rate was 33.7, corresponding to an annual average growth of national income of 11.3%. Then the growth rate diminished to an annual average rate of 7.2% during 1976-1980 and to 4.4% during 1980-1985, while the accumulation rate diminished also to 27.9%.

To put it in perspective, the real output per capita in Japan during 1950-2000 increased by 11.4 times, in the USA only by 2.6 times. These rates of growth were strongly correlated with the rate of technical progress, implying that technology could be a good way towards convergences.

Nowadays, countries like China, Hong Kong, Singapore, Taiwan, Malaysia, Thailand, Indonesia, and South Korea are fast growers. So, what are the secrets of growth, if there are such?

2. Long-term growth and short-run business fluctuations

In the period 1991-2002 the real GDP in Romania fluctuated around a slightly upward trend. In the long run the real GDP, in most economies, fluctuates around a rising trend. Long-term economic growth is reflected in the increasing trend of real GDP, irrespective of the business cycles, which are fluctuations of real GDP around its long-term trend.

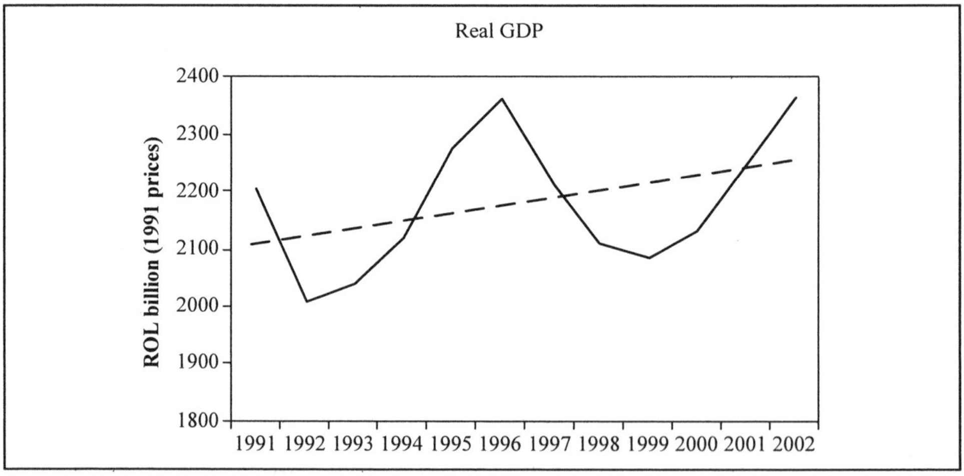

Since it is more difficult to calculate the potential GDP, we will consider the trend value of real GDP as an approximation for potential GDP. The chart above is based on a linear approximation of potential GDP.

Potential GDP measures what the economy could produce if all resources (land, labour and capital) were employed at their normal levels of utilization.

3. Factors of economic growth

The trend movement of real GDP is predominantly driven by the supply side of the economy. We will limit to three the main reasons why potential GDP grows: capital, labour input and productivity.

We will take into consideration the labour force defined as the number of persons who are either employed or unemployed, and the unemployment rate as the percentage of the labour force that is unemployed.

We will consider the volume of physical capital in any given year as it is determined by investment in previous years, and the investment as a flow of new capital during the year, that contributes to the changes in the stock of capital.

However, total factor productivity incorporates the productivity of both labour and capital, and the process of technological change tends to have a positive impact upon factor productivity.

Robert Solow of M.I.T., who received the Nobel Prize in 1987, developed a framework that can be used to determine the contributions of labour, capital, as well as total factor productivity to economic growth of a country (Solow, 1970).

His formula was used, among others, by Edward Denison of the Brookings Institution to study the contributions of each factor to long-term growth in the United States (Denison, 1974).

Assuming that GDP rate $\mathbf{r^y}$ equals productivity rate r^A plus the weighted rates of employment $\mathbf{r^L}$ and capital $\mathbf{r^K}$:

$$\text{then } r^y = r^A + a \cdot r^L + (1-a)r^K,$$

where the weight denoted "**a**" represents the share of labour cost in value added.

This growth formula can be used to infer the rate of total factor productivity as a residual:

$$r^A = r^y - a \cdot r^L - (1-a)r^K$$

3.1 Explaining Romanian economic growth

During the past decade, overall growth performance of the Romanian economy was disappointing.

Real GDP was highly volatile and its potential level grew at an extremely slow pace.

According to IMF estimations, potential GDP grew at an annual rate of 1.2 percent in the period 1992-2001.

The explanation of the low growth rate requires a closer look at employment, capital developments and total factor productivity.

Employment persistently declined throughout 1992-2001 at an annual rate of -2.0 percent.

Capital stock data is not available for Romania for that period, but IMF estimates suggest that it slightly increased by about 1 percent annually over the 1992–2001 period.

Table 1: Source of GDP growth in the period 1992-2001

			Annual percentage change
GDP			1.2
	Rate	Share Contribution	
Employment	-2.0	0.5	-1.0
Capital	1.0	0.5	0.5
Total factor productivity			1.7

4. Total factor productivity and growth policies in Romania

Based on the statistical evidence we assume a labour share of 0.5; this implies a modest contribution of capital (0.5 percentage points) to GDP growth.

At the same time the employment contribution was negative, estimated at 1 percentage point.

It emerges that total factor productivity, as measured by the Solow residual, provided the major contribution to GDP growth, which means 1.7%

So one can see that the positive rate of annual percentage change of GDP of 1.2% has resulted from a negative contribution of employment by –1%, by a positive contribution of capital by 0.5%, cumulated with a positive contribution of total factor productivity of 1.7%

Practice has shown that the government can influence all three of the determinants of growth: labour input, capital formation and total factor productivity.

The idea of stimulating productivity has been attractive to policy-makers, but there have been few proposals for concrete action.

Perhaps the most important role that the government can play in improving productivity growth is in the area of education and research and development.

A highly skilled labour force is obviously a key ingredient to successful productivity growth. In this respect Romania is facing a strange paradox: while the creative capacity is over average, the mechanism for using it is weak (Nastase, 1994).

On the other hand, increased growth in the capital stock requires consistently high levels of investment spending. Starting next year, tax cuts and investment tax credits may be utilized for stimulating investments. However, the experience of other economies in transition shows that the effects of the investment tax credits tend to be rather disappointing.

Tax cuts may also be used to improve work incentives. However, one should keep in mind that labour supply is rather inelastic, so that even a large tax cut has only a small effect on employment.

It is thought that lower taxes may provide incentives to work (the substitution effect), but the higher level of after-tax income generally depresses work (the income effect).

The government also plays a role in sponsoring research in science and engineering, and during the last 3-4 years a consistent policy in this field has been pursued in Romania (Chirovici, 2004).

4.1 The role of institutional factors

Economic growth also depends on factors that facilitate the production, the distribution and the sale of goods and services, which economists refer to as the economic infrastructure.

In this context, infrastructure does not refer to roads, structures, dams or other capital goods, but to institutional factors:

- Forms of government (encouraging private ownership and free enterprise, enforcing contracts and stamping out corruption, protecting physical and intellectual property);
- Monetary policy (ensuring the independence of the central bank committed to achieving price stability);
- Trade policy (supporting free trade);
- Regulatory framework (balancing the costs and the benefits of government intervention in the private sector).

During the first 10 years, after 1989, the stop and go policy in Romania was not conducive towards a harmonious social capability. It is only since 2000 that a consistent strategy has been developed in this respect.

4.2 Determinants of productivity growth in Romania 1992-2001

The hypothesis of the IMF is that the more efficient utilization of labour and capital was among the major determinants of total factor productivity growth at an annual rate of 1.7 percent throughout 1992-2001.

Privatization and restructuring improved, albeit rather slowly, the utilization and allocation of factor inputs, reducing hidden unemployment. At the same time, the new private sector, institutional reforms, more efficient management structures, as well as human capital accumulation in accordance with the requirements of a market economy gradually strengthened, sustaining the timid productivity gains.

5. Natural unemployment and the trend of Romanian unemployment rate

The institutional structure of the economy gives rise to a „natural" rate of unemployment toward which the economy gravitates, and which limits the efficient employment as factor of economic growth.

When the economy is at potential GDP, the unemployment is at its natural level and the factors of production are used at their normal intensity. Efforts to lower unemployment below the natural rate can succeed only in the short run and only by inducing upward pressures upon prices. Or in Romania, with relatively high inflation, there is no scope for such development (Dorbrescu, 2000).

However, estimates of the natural rate of unemployment provide a useful benchmark against which to gauge the current performance of the economy, as measured by the actual unemployment rate. Natural unemployment is said to occur when the only existing unemployment is frictional and structural.

The unemployment rate fluctuated around an upward trend suggesting that the rate of natural unemployment has increased in the period 1991-2002.

The increase of natural unemployment is not surprising in periods of intensive restructuring. If job creation is slow in compensating for job destruction, then the trend of unemployment is up.

But employment levels shrank by about 1.4 million people throughout 1991-2002.

In the same period, unemployment surged by more than 0.7 million people.

The labour force fell by about 0.7 million, moderating the effects of restructuring on unemployment.

Besides negative demographic trends, overly generous pension regulations played their role in fostering the decline of labour force, and consequently in employment.

There are expectations that the decline in the labour force might cease. Though continued restructuring could result in further layoffs, new industries and the service sector should be able to absorb these workers. Therefore, the unemployment rate is expected to stay around its current level of about 7-8 percent.

6. Economic growth, 2001-2004

Based on the same methodology we will continue our analysis for the period 2001-2004 (estimates for 2004), when the average growth rate has been around 5%.

Table 2: Source of GDP growth in the period 2001-2004

			Annual percentage change
GDP			5
	Rate	Share Contribution	
Employment	0	0.5	0
Capital	2	0.5	2
Total factor productivity			3

Taking into account the developments from the last 3-4 years, when real GDP has increased by about 5% per year, we notice from the above table that total factor productivity has contributed to the GDP growth by 3 percentage points, reflecting a strong component of innovation activity, since the more efficient reallocation of labour and capital registered in the previous period has been approaching normality. And this has come mainly from the demand side due to the "reanimation" of industrial activity. This is confirmed also by the fact that foreign investments in Romania have been concentrated (64.7% at end of March 2004) in industry, at the same time taking advantage of and bringing innovation to Romania.

On the other hand, the factor capital has contributed to the GDP growth by 2 percentage points, reflecting the fact that the investment ratio to GDP has increased from around 17% in 1999 to 22.5% in 2003, and the trend is an increasing one.

The correlation between industrial production and innovation developments has been quite strong. Going back, in 1992 industrial production was less than in 1989 by 37.3%; it increased by almost 18% between 1993-1996, only to decrease again by 14% between 1997-1999.

Starting with 2000, it has increased by more than 7% on average, being the main driving force of the GDP since then. It is expected that this year industrial production will reach the 1990 level.

The research and development picture has reflected, in general, but with a certain delay, the fluctuation of industrial activity, showing the strong importance of the demand side for this type of activity during the transition period. Only relatively recently have the economic agents rediscovered the utility of their own research activity: the fact that they cannot depend on imports in this domain has become clear for them, and they abandoned the idea that everything should come from abroad. Inside research (factory research) has started to yield important results, and new industrial sectors (science-intensive) have appeared.

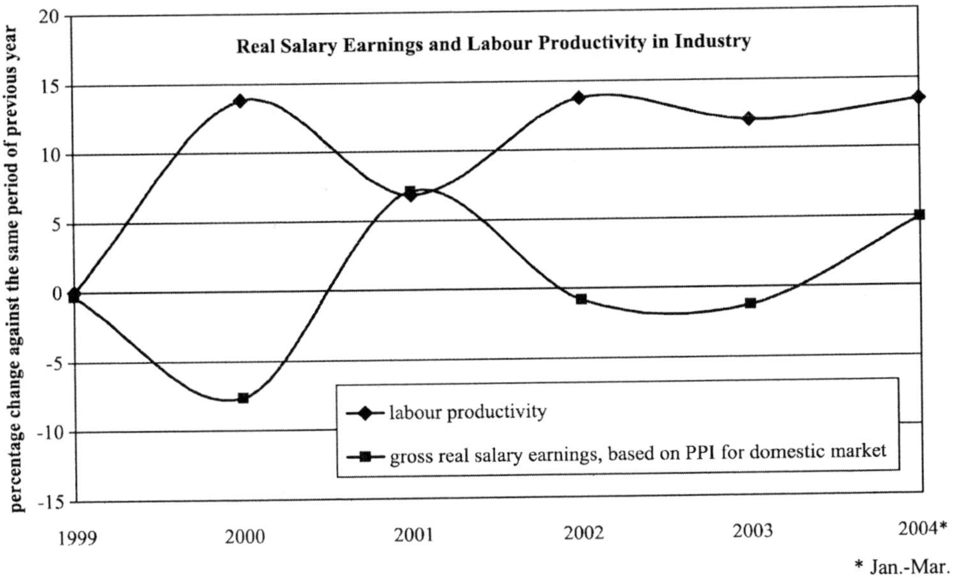

Fig. Real Salary Earnings and Labour Productivity in Industry
Source: National Institute of Statistics, NBR calculations

Starting with 2000, as one can see from the above graph, industrial productivity has gained a sustainable trend, based on investments and innovation.

This trend, if continued, resembles the trends registered in France during the 50s and 60s, and could be conducive to steadily decreasing the gap between Romania and the developed countries. Already the GDP per capita has increased from 6000 Euros to an estimated 7250 this year.

In spite of all these efforts, only this year will the Romanian economy reach the 1990 level, although in certain fields it has made significant progress, which could represent a necessary shortcut for the catch-up process.

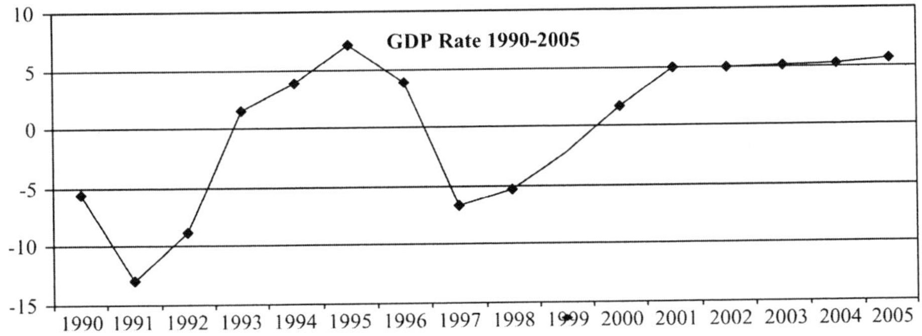

Fig. GDP Rate 1990-2005

7. Conclusion

The results of this short analysis show that innovation, through the total factor productivity, played a determinant role in the economic growth in Romania during the transition period.

Although during the first 10-11 years of the post-socialist transition the annual average rate of growth in Romania was very modest, 1.2%, the total factor productivity contributed with 1.7 percentage points, the majority of which came from a reallocation of labour and capital.

But during the last 4 years, in spite of the fact that the reallocation effect has diminished substantially as privatization has advanced in a predominant way, the total factor productivity has increased its importance to 3 percentage points. This is coherent however with the development of new products and the introduction of new technologies in the economy. The question remains if one can consider this as a sustainable innovation and if European Union funding for innovation will be conducive towards long-term economic growth and competitiveness (Leal Filho, 2004).

References

Chirovici, E.O.(ed.) (2004) Proprietatea intelectuala in activitatea intreprinderilor mici si mijlocii, Editura Global, Bucuresti.

Denison, E (1974) Accounting for United States Economic Growth 1929-1969, Brookings Institution.

Dobrescu, E. (1979) The optimum of the socialist economy, Editura Academiei Republicii Socialiste Romania, Bucuresti.

Dobrescu, E. (2000) Macromodels of the Romanian transition economy, third edition, The Expert publishing House, Bucharest.

Leal Filho, W. (2004) Trends in innovation in a European context-problems and perspectives. In Leal Filho, W. (ed.) Supporting the development of R&D and innovation potential of post-socialist countries, IOS Press, Amsterdam.

Nastase, G. (1994) Probleme globale ale potentialului inovator, Bucuresti.

Postolache, T. (ed.) (1991) Economia Romaniei-secolul XX-, Editura Academiei, Bucuresti.

Romer, P.M. (1989) Endogenous Technical Change, NBER Working Paper No.3210, Cambridge Massachusetts.

Solow, R.M. (1970) Growth Theory, Oxford University Press.

Chapter 14
The Innovation in Education: the Past, the Present and the Future

S. ZENCHANKA, S. MALCHENKO, A. JAKIMAHO
Moscow State University of Economics
Statistics and Informatics
18a, Zheleznodorozhnaya Str.
220014 Minsk
Belarus

Summary

The development of the education system, the creation of the all-European education system, demands from the different countries joint efforts in development of the new syllabuses and curricula. This approach demands to define possible ways to Bologna process as for countries that joined to this process as for countries that plan to do this.

Introduction

At the modern condition of globalization and efforts of the different countries in their moving to sustainable development, higher schools and education systems as a whole must define the path of future development.

It is necessary to mention that education has two different sides – from the one it is a very conservative area of activity, based on extant knowledge, while from the other it must be based on mostly modern scientific and methodical developments. These two sides of the education enter into collision, and the role of innovation role is to confront this dichotomy.

The national aspect of education is an important side of the education process: it is impossible to demand the coincidence of all educational components in countries with a different geographic position, development level, varying culture and so on.

There are some supranational aspects of education in different branches of knowledge that must be realized in the context of education activity, independently of the country.

The Bologna Declaration (Joint Declaration, 1999) signed in 1999 by the Ministers of Education of 29 European countries defines total tasks for the creation of the common European area of education to the year 2010. At the Prague conference (Communique, 2003) this obligation was confirmed and the task quantity was increased.

In the Berlin conference decisions (Communique, 2003) the importance of all elements of the Bologna process was underlined and the need for effort activation was picked out at the university, national and European levels. At this conference the importance of the education quality guarantee, the acceptance of a two level (bachelor and master) education system and student and teachers mobility was pointed out and so on.

The obligation that higher education be made accessible for everyone is a very important element of the Bologna process. At this conference the quantity of countries that signed the Bologna Declaration was increased to 40. Of the former Soviet Union countries, Lithuania, Latvia and Estonia joined this process in 1999, and Russia in 2003. Hence the Bologna process is an innovation process for joined countries and its starting stage must be finished by 2005.

What innovation priorities must be selected?

- The introduction of new syllabuses and curricula for specialities and specializations (integrated curricula);
- The development of a new methodical and methodological approach to education (one aspect is the interactivity of education, which must lead to a decrease in lecture attendance by students);
- The using of the modern scientific achievements and participation of the teachers in scientific research;
- The introduction of new technical educational methods;
- Open education.

We will consider three main periods:

1. The Last (till 1992) – the Soviet Union period, which had a recognized education system.
2. The Present – (from 1992 till the present time) – the adaptation period of the education systems in former Soviet Union states to new conditions.
3. The Future – the development of education systems in accordance with the Bologna Process.

The past

The education system of the Soviet Union was a state system, with the distribution of graduates to different enterprises and organizations. The graduates had to work at these enterprises/organizations for three years, and after this time they could choose their new place of work. The system guaranteed work to graduates, and promoted a constant influx of young specialists to all branches of the economy.

Some higher schools were oriented to specialist training for specific branches of industry and science, and leading specialists from these branches took part in the educational process. Besides this there was a system of practical work experience, whereby over several years the students took part in the industrial process at various enterprises or at scientific laboratories. This helped them to a deeper mastering of the syllabus and curricula.

It is necessary to consider the system of finance support for education (i.e. investment). It was pointed out that there was a strong connection between enterprises (organizations), science and higher schools. Now we consider the investment process into these branches.

The first stage of the investment process was a Soviet Union budget development. This budget defined finance measures for different branches of the economy. There were special items for the financing of education, science and different branches of the industry. It is necessary to indicate that the budget financing of science means fundamental science financing. Such investigations were conducted at the Academy of Science institutes and leading universities. The applied investigations were carried out basically using other budget items. Such a budget structure allowed higher schools to get to scientific investigation as the direct budget financing as indirect financing from academy institutes and industry. Hence the Soviet Union education system had close connections with industry and academy institutes and conformed with the demands of science.

The education systems of the Eastern European countries corresponded in many respects with the Soviet Union education system, and thus this system was a transnational one. At the same time it had a number of deficiencies:

1. The graduates were localised in the place where they chose to work. The results of the distribution were connected not only with graduate knowledge but with their public activity. Graduates must go to the new towns and they could not often go back to their native town. On the other hand there was a personal rotation as in a separate ministry as between different ministries. Mostly successful graduates even from off-centre higher schools could settle at ministry.
2. The education was rather conservative especially for fundamental courses. Syllabuses were overloaded with theoretical disciplines and this knowledge did not often find application in practical use. That is why graduates could not find their place in the economy and must search for new area activity.

The Present

In December of 1991 and January of 1992 the agreements and protocols were signed that defined the principles and features of a new interaction between former Soviet Union republics – the Commonwealth of Independent States (CIS). In 1992 all these republics made Independent Declarations. As a result the communications in different areas were destroyed. All New Independent States (NIS) must start from zero level because in the Soviet Union all significant problems were solved in Moscow. This process was preceded by the disintegration of the Warsaw Pact and the exit of the Eastern European countries from the Soviet Union area of influence.

The economy of all NIS including Russia came down. The budgets of NIS had no money for scientific research development and education. This led to a reduction in teachers' and scientists' salaries. A specialist drain from educational and scientific organizations set in: the number of young people who wanted to work at scientific and educational organizations decreased. As a result the average age of teachers and scientists increased.

This had an effect on the quality of the education process and scientific research, because to maintain adequate living standards teachers and researchers had to cope with a bigger volume of work. Teachers could not take part in scientific investigations and researchers could not take part in the education process. As a result a new problem appeared – aged professors do not possess modern methodical and technical means of education and do not possess modern knowledge, which leads not only to impairment of the quality of the teaching process but to decreasing education quality as well.

Legislative base of education

In the Law of the Republic of Belarus "About education" (Law, 2002), some fundamental definitions are made. This law defines all aspects of the educational process at all levels – secondary education, specialized education, higher education. For higher education two levels (bachelor and master) are defined. There are several different possibilities of getting into higher education:

- full-time instruction (daily and evening training);
- extramural instruction;
- self-sufficient instruction;
- individual curriculum.

It is possible to combine different form of education

Competition in education

State higher education includes 44 higher schools of different types (university, academy, institute and others). More than half (23 higher schools) are situated in Minsk and rather – at 5 district cities and 3 big region centres . There are 13 private higher schools in Belarus (8 from them are in Minsk).

There are 2 international higher schools and two branches of Moscow universities.

Numbers of students increased from 189 thousands in 1988 to 320 thousands in 2003. *272.9* thousands of students (85%) prefer to study at state institutes. The rest (15%) get education at private institutes.

Hence the Belarus student prefers to get higher education at state institutes and free of charge.

Demographic situation

There is a demographic decline at the Republic of Belarus. We have a decreasing of young people aged under 17. This can lead to a reduction in the numbers of secondary school graduates who will enter higher school. A similar situation exists in Russia.

The Future

An analysis of the legislative base and demographic situation shows that it is necessary to change the approach to the process of higher education. As world experience shows, a significant enhancement of the education process for training and retraining can be reached via implementation of the remote education system. In accordance with world standards the modern remote education systems are based on INTERNET technology.

As mentioned above, the elements of the Belarus law "About education" coincide with the Bologna Declaration in part of the education levels. This gives the possibility for the Belarus education system to join the Bologna process in the future. Now it is necessary to develop the structure of curricula and syllabus for compatible and comparable qualifications.

To guarantee the quality of the education system it is necessary to introduce in higher schools quality management systems, in accordance with ISO 9001 or TQM principles (ISO 9001).

One of the most important problems is a guarantee of the university independence and autonomy. These rights are defined in legal terms, but in practice we have a state system with a strong hierarchy.

There is the limited university staff and student mobility. For a student it may be explained by absence of the compatible and comparable curricula and syllabus. Higher school graduates sometime get a possibility to continue their education as a Master or PhD student at European or American universities.

It is finally necessary to develop a system of long life learning that helps people to raise their education level, get a new speciality and progress in their work.

The Minsk branch of the Moscow State University of Economics, Statistics and Informatics started its activity in 2003. From the beginning this branch has tried to adhere to the Bologna principles, because Russia has joined this process. It has a high level of informatization:

- round-the-clock access to INTERNET;
- round-the-clock access to e-mail;
- an electronic library.

In the area of open education the branch is a provider in Belarus of the Russian system of *externat*.

In the area of lifetime learning, the branch offers people various different retraining courses and courses for the improvement of their qualifications.

Conclusion

The state of the system of higher education, ways to improve this system and its progress in the direction of the Bologna process have been considered. The positive and negative sides of the Soviet Union education system have been shown. The factors that can have an influence on the development of the educational process have been defined.

References

The Communique of the European Ministers of Education, Berlin, 19 September 2003, www.umo.msu.ru/cjnf/berlin.htm

ISO 9001:2000. Quality Management Systems. Requirements.

Joint Declaration of the European Ministers of Education "The European Higher Education Area", www.ntb.ch/SEFI/bolognadec.html

Law of the Republic of Belarus "About Education". National roll. – 2002. – N 37 2/844.

Chapter 15
An Overview of the Role of Funding and Other Mechanisms Supporting Innovation

WALTER LEAL FILHO
TuTech Innovation
Harburger Schlossstrasse 6-10
21079 Hamburg
Germany

Summary

This chapter presents an analysis of the role played by financing in innovation and describes a set of mechanisms which may used to support innovation in Europe.

Introduction

Many publications have looked at the factors influencing the way technology is managed (e.g. Betz, 2003), addressed how technological innovation in general occurs (Christensen 2003, Thomke 2003), discussed concepts and techniques for managing innovation in Europe (Leal Filho 2002, Leal Filho 2004) and examined specific challenges in areas such as the environment in the Baltic region (e.g. Leal Filho and Ubelis 2004). In addition, there are sets of publications dealing with the day-to-day aspects of innovation such as "Innovation in Europe – results for the EU, Iceland and Norway – data 1998-2001" (Commission of the European Communities, 2004) which have tried to benchmark the provisions for innovation in different countries.

Yet few publications have addressed ways in which funding and financing can support innovation. Indeed, from a managerial perspective, there is a lack of specific works which may provide a strategic view of innovation and demonstrate that it is more than just encouraging creativity. The reasons for this gap are many. Innovators and entrepreneurs do not often engage in discussions on financial issues. Indeed, such discussions are not only avoided, but also deliberately passed on to others, since the general impression is that an innovator is not usually competent enough to handle financial and technical matters at the same time. One things is clear: the matter is complex but needs to be tackled.

Owing to the high degree of risk and uncertainty, funding to innovation is not the same as in other categories of investment and, as such, investments in innovation infrastructure do not get treated the same way as investment in other fields of the economy. In addition, the technology industry in Europe as a whole – and in southern Europe in particular – has, with a few exceptions, experienced relatively low levels of demand for both grants and business loans for their innovation efforts. Yet the need for business capital remains an important issue. Therefore, the demand for financing in the technological field could be increased if grants and other financial products and delivery systems were better designed to meet the particular demands and characteristics of scientists and entrepreneurs.

The innovation capacity of companies is acknowledged as being one of the factors that most contribute to the competitiveness of a country or region. It is also a key component in economic development, long-term growth and employment. Therefore, it is necessary to foster in companies a culture that allows them to benefit from the opportunities derived from financing which may jump-start significant and even modest technological changes. Having said that, one of the most important obstacles that innovation encounters is the restraint of financing due to the high-risk component of the investments. Consequently, it is necessary to articulate measures that support financing innovation activities at regional, national and European levels, an issue which will be explored in detail in this chapter.

Funding for what?

A fundamental question to be answered when analysing the role of funding and other support mechanisms for innovation is the purpose of that funding or, in other words, one can ask the question "funding for what"? As a whole, innovation-related funding is available for:

- development of a product
- optimisation of a process
- development of scientific and technical methods
- production of equipment
- help and support to SMEs in traditional or new areas, by boosting their technological capacities.

When one considers the issues which need to be considered to attract funding to implement innovation, one may suddenly realise that creativity and innovation are not the same thing. In most companies, there are many good creative ideas that never move into the action phase, and, therefore, do not become innovations nor receive any funding. Both creativity and innovation are very relevant to strategic management. As stated in Hussen's book "The Innovation Challenge" (1997):

> "It is the ability to innovate that determines much of what an organization is able to do. Innovative capability provides the organization with bricks from which a strategy can be formed and which can eventually evolve into a bastion of strong competitive advantage."

Unfortunately, funding possibilities for policy support, training of staff or marketing are not normally included in innovation funding packages. This is a negative trend in the sense that even the best products and processes will not be useful if the staff are not trained and if there are no ways of making them known to the target groups or audiences. There is a therefore a need to enlarge the view of innovation funding, also in order to cater for the support needed to bring a given product to the market place. Figure 1 illustrates some areas of funding activities in the EU and some in Germany.

It is noticeable that the variety of tasks which may receive funding support is significant. But at the same time Figure 1 illustrates the fact that, although there are some overlaps in certain areas, funding at the national level may compensate for the deficiencies seen in respect of European funding.

It is often the case that funding for innovation becomes easier to obtain if collaborative efforts between research organisations and industry take place where such interaction is of benefit in its own right and also assists knowledge transfer. In doing so, a range of additional funding mechanisms may become available, which may ultimately lead to the desired funding and, to a great extent, to increased chances of the pursued innovation activity yielding the expected benefits.

European Funding Activities	German Funding Activities
Cross-cutting Research Activities	
• Research for policy support • New and emerging science and technology (NEST) • JRC activities	• Collaborative research at large-scale facilities • Technologies in the humanities and mathematics
Joint actions	
• Co-ordination of research activities • Development of research/innovation policies • Specific SME activities • International co-operation activities	• InnoRegio • Innovative regional core growth areas • University-based start-ups (Exist)
Strengthening Research Structure	
• Research and innovation • Research infrastrutures • Science and society	• IT information services • Innovative work organisation • Development of competence • Equal opportunities for women in education and research • Invention and patent support
Human Resources and Mobility	**Research in Higher Education**
• Marie Curie Actions	• Programme supporting application-oriented research and development at universities of applied sciences (aFuE) • Funding by the DFG • Postgraduate study groups • University construction including large-scale equipment and communication networks

Fig. 1: Areas of funding activities in the EU and Germany

Some current funding mechanisms

There are a number of mechanisms which may support innovation in European countries today. The European Commission, being aware of the contribution that innovation can make to driving up productivity and to achieving the vision of Europe being a key knowledge hub in the global economy, supports innovation in a variety of ways. Building on progress made in previous innovation efforts, we have today various different means via which businesses may be encouraged to increase their investments in new products and services with the aim of continuously increasing the added value and thus maximising their innovation potential.

The INNOVATION/SMEs programme, for example, is managed by the Innovation Directorate of DG Enterprise, has a legal mandate under the Council Decision on the 5th Framework Programme and its Work Programme to promote the exploitation of RTD results as well as the emergence of a favourable environment for financing the exploitation, adaptation and dissemination of technology. This includes:

1. the interlinking of financiers, investors and technology sources
2. initiatives to mobilise private capital for the financing of innovation

It is relevant at this stage to highlight a number of activities organised in this field by the Innovation directorate of DG Enterprise.

a) Innovation and Technology Equity Capital (I-TEC)

Launched in July 1997 in cooperation with the European Investment Fund, I-TEC aims to foster investment in early stage, innovative technology projects by private Venture Capital operators. The I-TEC Network of Venture Capital Funds gathers funds, which have committed to invest at least 25% of their capital in such projects.

b) Support to the establishment of specialised networks of financiers and investors

Since 1984, the Innovation directorate of DG Enterprise has supported actively the establishment and development of the European Venture Capital Association (EVCA), which currently gathers more than 400 professional Venture Capital operators. The Innovation directorate of DG Enterprise represents the European Commission on the Board of the EVCA.

From 1994 on, the Innovation directorate of DG Enterprise supported the creation of the European Association of Securities Dealers (EASD), today representing some 75 investment banks and market professionals. New Stock Markets for High Growth, Innovative Companies such as EASDAQ and EuroNM find their origins in these initiatives.

c) Transnational Investment Fora

Transnational Investment Fora bring together innovative SMEs and potential investors. 16 such fora have been organised across Europe, in cooperation with the European Venture Capital Association and with national innovation agencies, involving more than 500 innovative companies. The concept has been integrated in Community RTD programmes (e.g. Esprit, Brite-Euram), and additional events are planned.

d) The FIT Policy Exchange Project

The FIT Policy Exchange Project aims to organise a systematic exchange of good practice in Financing Innovation between Member States' policy makers and market operators. Six subjects (expanding) are to be covered in the coming months, with workshops and dissemination activities.

e) Training Seminars on Financing Innovation

The INNOVATION-SMEs programme organises training seminars for project officers and participants in Community RTD programmes. These seminars are given by a venture capitalist, a banker, a high-tech entrepreneur. More than 150 participants have already attended these seminars, and further events are planned.

In addition, there are various activities aimed at promoting the exploitation and financing of RTD results. One of them is the "The Gate2Growth Initiative".

The prime objective of the "Gate2Growth Initiative" is to support innovative entrepreneurs in Europe. In addition, Gate2Growth aims to assist service providers improve their capacity to assist entrepreneurs by fostering networking and the exchange of experience and good practice at European level. These include early stage technology venture capital investors, managers of technology incubators, managers of industrial liaison and technology transfer offices linked to universities and research centres , academics in entrepreneurship, finance research and teaching. Gate2Growth thus provides tools, infrastructure and support services directed to innovative entrepreneurs as well as to their supporters.

"The Gate2Growth Initiative" was complemented by LIFT. The LIFT pilot project provided a helpdesk on financing of innovation to support and direct participants in Community RTD programmes in the exploitation of results, as well as to assist thematic RTD programmes in establishing links with financiers and investors. LIFT was operational from mid-1999 to the end of 2001. LIFT functions have been extended in the Gate2Growth Initiative.

The Gate2Growth Initiative is part of the Innovation/SMEs Programme action line "access to private innovation financing and tools for better knowledge exploitation", managed by the Innovation Policy Unit of DG Enterprise. It incorporates the lessons learned from, and consolidates the worthwhile elements from the following pre-cursor pilot projects:

- the I-TEC pilot project, launched in 1997 in cooperation with the European Investment Fund
- the LIFT helpdesk, launched in 1998
- the FIT exchange projects on technology incubation, entrepreneurship training, informal investors, guarantee mechanisms, technology rating and finance academia.

The ultimate objective of the Gate2Growth Initiative is to support the innovative entrepreneurs of Europe.

In addition, the initiative focuses on helping important support players to improve their capacity to assist entrepreneurs by networking and exchanging experience and good practice at European level. These include:

1. early stage technology venture capital investors,
2. managers of technology incubators,
3. managers of industrial liaison and technology transfer offices linked to universities and research centres,
4. academics in entrepreneurship, finance research and teaching.

Gate2Growth thus provides tools, infrastructure and support services directed to innovative entrepreneurs as well as to their supporters. For technology entrepreneurs and participants in Community funded RTD Gate2Growth.com, a one-stop-shop for the technology entrepreneur, giving access to:

- Business plan preparation tool package
- Business plan diagnostic
- Discussion forums
- Industry news and events
- Investor identification and matching tool
- Service provider directory
- Seminars and workshops
- Peer group and entrepreneur clubs
- Access to a network of local intermediaries.

Furthermore, there are other broader support mechanisms for innovation professionals and intermediaries in particular incubator managers, ILO and IPR managers and to early stage technology venture capital investors. These are:

- Industry training voucher
- Pan-European secondment programme
- High tech cluster visits (including USA)
- Access to experts, trainers and speakers (including USA)
- Industry conference access

In respect of community and networking, the following programmes are being offered:

- I-TecNet: pan-European network of 70 early stage technology venture capital investors
- G2G Incubator Forum: pan-European network of 150 technology incubators
- ProTon Europe: pan-European network of 250 ILO managers
- G2G Finance Academia: pan-European network of academics and entrepreneurship trainers
- Annual Gate2Growth partnering event
- Gate2Growth.com thematic homepage.

Moreover, in terms of tools and resources, a set of services is also available, including:

- Investor and technology matching service
- Online membership and compentence directory
- A support package for entrepreneurship training providers
- A support package for Business Plan Contest Organisers

Of further interest to those interested in the funding of innovation is the "TEC: Innovation and Technology Equity Capital". The I-TEC pilot project is an initiative to encourage early stage investments in technologically innovative SMEs. It was launched in July 1997. I-TEC is supported by the European Commission under its INNOVATION/SMEs Programme, and implemented in collaboration with the European Investment Fund.

I-TEC aspires to help build, within Venture Capital operators, a lasting capability to appraise and manage those early stage projects, in technologically innovative SMEs, which, in spite of their economic viability and inherent quality, would otherwise not be taken into account.

I-TEC will benefit European innovative SMEs, active in the European Union and in the Associated States (i.e. Iceland, Israel, Liechtenstein, Norway), which should directly approach the selected I-TEC Venture Capital operators.

Thanks to the I-TEC pilot project, innovative SMEs can access a network of 28 capable Venture Capital investors, interested in business projects with a high degree of innovation in technology, product, service or process and which exhibit a potential for high growth and new job creation.

The Venture Capital operators which are taking part in the I-TEC pilot project have agreed to devote at least 25% of their funds to early stage investment in technologically innovative SMEs over the next years.

Finally, mention should be made of *bank debt*. Contrary to popular belief, banks do lend to innovative companies, but generally only where repayment sources have been identified, which typically means that the company needs to be trading, and that sales need to have been identified. When looking at bank debt as a route to fund innovation, certain things need to be considered:

- it is short term
- banks are averse to risk
- the money is repayable now
- a positive cash flow is required
- You could consider using a bank for:
- working capital
- financing raw materials

- hiring staff
- funding ideas which are close to being manufactured or sold.

In addition to loans or overdrafts, there are other important financial instruments, such as: asset finance for lease or purchase of machinery invoice finance to release cash flow from book debt credit cards.

The primary attraction of debt is that it avoids the dilution of ownership, which may be an important issue, depending on the future growth potential of a given innovation. However, debt is not the best way for an early company to finance itself. In many cases, the need to be investment-ready is primary, which means having 'the ability to give confidence to the prospective funder that the business will be able to repay any proposed borrowing, or to provide a suitable return on any likely investment'. Evidence of the company's investment readiness will include a strong management capability with:

- a well developed product and/or service
- good market knowledge
- personal commitment to the success of the opportunity
- business plan

It is often the case that problems arise in one or more of the above areas. In such circumstances the innovator is best advised to try other sources.

The United Kingdom as an example

One of the prime examples of the variety of support given to innovation comes from the United Kingdom. For this reason, it appears appropriate to try to describe some of the various funding means and mechanisms available there, in the hope that they may serve as inspiration to those countries or regions where support for innovation is at present rather limited. Some initiatives supported by the UK public sector include:

- central Government initiatives
- European Union initiatives
- RDA and local authority grants
- business support agencies
- support to universities.

The first example is the programme "Business Link". It develops schemes to assist companies to access the appropriate finance. The Access to Finance programme in the Objective Two regions of London, for example, offers assistance to organisations seeking to raise finance, in order to develop a marketing or business plan. It undertakes a free funding review, and offers subsidised support of up to 75% of that activity for existing businesses, and up to 87.5% for start-ups or pre-starts.

A second examples is from the UK Inland Revenue. It manages a national programme that assists companies and individuals with a specific innovative idea, offering up to a 75% subsidy of specialist advice and project management, up to a total of £12,000. The firm or individual must provide at least two days of work for each day of consultancy during the project, and the output will be a detailed action plan on how that product or service might best be brought to the market. R&D grants are available to SMEs to develop technologically innovative products and processes. The three main areas of support for that programme are:

up to 50% of a £20,000 project for micro-companies, a 60% subsidy on projects of up to £75,000 for small companies and up to 35% subsidy for medium companies.

Tax credits are also available from the Inland Revenue for R&D, and allow up to 150% of qualifying expenditure for the programme. It is vital to ensure that projects fall within the correct definition of R&D, and it can be offset against tax. There are several knowledge transfer partnerships providing access to higher education research, or funding for graduates to work on your R&D programme.

A further example of support is the "NESTA's Invention and Innovation programme" which was set up to fund ideas from their very earliest stage. NESTA is regarded as a creative investor, which gives it great flexibility in the ideas that they can fund, looking for a mixture of social, environmental and cultural benefits. NESTA invests up to £250,000, and works with firms to build a team from a pool of mentors. It has an application process regarded as simple and quick, particularly in the early stages, and aims to give a first response within six weeks of an initial proposal. NESTA focuses on:

- thorough research of the competition
- evidence of patent searches
- consideration of the route to market
- recognition of the need to build a team around the idea
- business plan for the duration of the funding.

The statement provided by an entrepreneur funded by NESTA exemplifies the process:

> In 1972, I received my first venture capital and began my business, and five years later, it employed 50 people. I started a second business, called Venture Capital Report, a monthly magazine for businesses seeking capital. By 1983, it became clear that scientists and technologists were finding it difficult to raise capital, due to the high risks associated, and the first fund I set up was for £123,000, which was invested in five projects.
>
> I am currently on my fifth fund, and my current funds are called the Oxford Technology Venture Capital Trusts. Investments in a venture capital trust attract a £2,000 tax rebate, and the first £4,000 of capital gains tax is also not payable. All dividends are free of tax, and sales of shares do not attract any capital gains tax. In return, the venture capital trust has to invest less than £1 million in unquoted companies. We have 900,000 individual investors in our three venture capital trusts.

At the national level, a further set of sources of funding for innovation are available in the UK. The following ones may be mentioned:

Research & Development: A DTi (previously SMART, which closed for applications on 31/08/03) scheme, which provides grants to help individuals and small and medium sized businesses (SMEs) to make better use of technology and to develop technologically innovative products and processes. Web: www.sbs.gov.uk

Grant for Investigating an Innovative Idea: Potential help for businesses who have an idea to develop an innovative product or service. Web: www.sbs.gov.uk

Small Firms Loan Guarantee Scheme (SFLGS): Small Business Services (SBS) scheme to guarantee loans from the banks and other financial institutions for small firms that have viable business proposals but who have tried and failed to get a conventional loan because of a lack of security. Web: www.sbs.gov.uk

Regional Venture Capital Fund (RVCF): Aimed at the "equity gap" i.e. equity finance above £50,000 and below £250,000 for growing SMEs, a fund of £25,000,000 is available. Web: www.southwestventures.co.uk

Foresight: Brings together key people, knowledge and ideas to look beyond normal commercial time horizons to identify potential opportunities from new science and technologies and actions needed to help realise those opportunities. Web: www.foresight.gov.uk

LINK: Aims to enhance the competitiveness of UK industry, stimulating innovation – providing support for managed programmes of pre-competitive science and technology in market or technology sectors. Encouraging industry to invest in further work leading to commercially successful products, processes, systems and services. Small and medium-sized enterprises (SMEs) are particularly encouraged to get involved. Web: www.ost.gov.uk/link

NESTA: Set up to support and promote talent, innovation and creativity in the fields of science, technology and the arts. Looking at filling the funding gap by investing in outstanding ideas at an early stage. Web: www.nesta.org.uk

These schemes are followed by awards to help innovation, such as:

a) *Global Watch Secondments*: DTI-sponsored programme to promote international staff secondments from 3-12 months. www.globalwatchonline.com/secondments/
b) *Knowledge Transfer Partnerships*: A government-funded scheme that enables businesses to access the skills and resources of high quality graduates. Placements can now range from 12-36 months. (Previously TCS). www.ktponline.org.uk
c) *International Technology Promoters Programme*: Offers access to a network of technology transfer specialists (ITPs) providing a practical route to leading-edge technological developments from the world's major investors in research and development. www.itp.org.uk
d) *Process and Marketing Grant (PMG)*: Enables capital investment in the processing and marketing of English primary agricultural products. Encouraging farmers to become more competitive, respond to consumer demand, be more innovative and add value to their products. http://www.defra.gov.uk/erdp/schemes/pmg/default.htm
e) *National Business Angels*: Network bringing together investors with commercial experience and companies looking for development capital. www.nationalbusangels.co.uk
f) *South West Business Angel and Investor Network (SWAIN)*: SWAIN is the Regional Business Angel network for the South West. SWAIN is a recently formed business introduction service which will bring together investors and companies seeking equity finance for growth. www.swain.org.uk
g) *British Venture Capital Association*: Sources of funds and professional advice. www.bvca.co.uk
h) *STEP*: Shell Technology Enterprise Programme, designed to help SMEs develop their potential by utilising the skills of undergraduates at affordable cost. Aimed at businesses that do not need a placement for 2 years, these can run between 4 to 12 weeks, mainly in the summer but also at any time of the year. www.step.org/
i) *Queen's Award*: The UK's top awards for business performance. www.queensawards.org.uk
j) *Small Business Service – Finance*: This link provides a look at where businesses can get funding, how to find out more about venture capital and what is available in the form of grants, funds and loan guarantee schemes. www.sbs.gov.uk/finance
k) *DTI Innovation Unit*: The Unit provides a host of useful tools for innovative businesses. http://www.innovation.gov.uk/

Many UK schemes also link with EU initiatives such as *EUREKA*, a pan-European network for market-oriented, industrial R&D which supports the competitiveness of European companies through international collaboration, in creating links and networks of innovation. The objective is to bring high quality research and development efforts to the market and to use the multiplying effects of cooperation.

How to absorb funding for innovation

The purpose of this section is not to turn scientists and engineers into financial experts, but to explain some of the principles behind the investment process and specifically, the meaning of the terms which are often used when discussing the roles of banks and other financial institutions in providing finance to companies.

As with many specialisms, finance and accountancy are full of jargon incomprehensible to the layman and scientists and innovation entrepreneurs are no exception. However, understanding these concepts, and why they are important, is essential to the understanding of the factors which influence providers of capital to companies, especially to new, technologically oriented firms.

Anyone interested in a company, whether as a supplier checking out its creditworthiness, a banker or a potential investor will be confronted by financial statements prepared by an accountant. These statements will generally include a balance sheet, a profit and loss account and a cash flow, among other documents.

On any particular date, a company's affairs should be in a state of balance, which can be represented by the simple equation:

Assets = Liabilities + Owners' Equity

In order to assist innovators with the complexities of finance and accounting, a project entitled "FISBI – Financing Innovation in Science Based Industries" has been started, with funding provided by the European Commission. The project, coordinated by Marco Da Rin from the Fondazione ENI Enrico Mattei (FEEM) in Milan, consisted of a series of three conferences (held between 1998 and 2001) and was aimed at creating a long-lasting network of financial and industrial economists interested in studying how financial markets affect corporate strategies in science-based industries. Its purpose was to gather young and experienced European researchers, working on both theoretical and empirical issues to create a specific network on the topic.

The three conferences created the opportunity to put in touch established scholars and young profession members, including graduate students. It is very common for financial and industrial economists to explore these issues within the scope of their own field, with very little 'cross-field' interaction. The project was intended to contribute to overcoming this major weakness. Young researchers from Eastern Europe were also involved. The first conference was held in Mannheim at the Zentrum for Europäische Wirtschaftforschung on January 20-21 1999, focusing on the specific topic "Financing entrepreneurship and innovation in science-based industries". The programme comprised papers from different backgrounds, with different methodologies and intended for different audiences. This ensured an active discussion and the involvement of several practitioners who attended the event.

The second conference was held at FEEM on 21-22 May 1999 covering the topic of the "Influence of financial markets on industrial restructuring in science-based industries". This conference was focused on a more technical theme, but papers which were related to the general theme of the network were also accepted in order to ensure continuity in the group of researchers attending the FISBI conferences. Presenters from three different groups were present: theorists, empiricists and management scientists. Nevertheless this mix turned out to be particularly successful.

The third and final conference was held at the Universidad Carlo III de Madrid in Madrid on 20-21May 2000 on the theme: "Financing R&D in science-based industries: a comparative analysis at the European level". It considered theoretical and empirical aspects of the financing of R&D activity with particular reference to the European experience. Most of the papers presented (6 out of 9) were devoted to empirical analysis with the remaining investing in theoretical aspects. The empirical papers covered both the experiences of individual countries (in particular Germany) and comparisons between European and non-European countries, while the theoretical papers focused on the optimal way to finance new firms. The mix of empirical and theoretical analysis helped in providing a lively discussion and in making participants aware of the work done in other areas.

The lesson learned from the project, which should indeed be reconvened and extended, is that only by means of specific training and networking can we make sure that innovators are able fully to absorb the level of funding available to them today.

Conclusions

This chapter has outlined the role of financing for innovation and has described some mechanisms via which it may be realised, with some concrete examples of what has been achieved. In concluding this chapter, mention ought to be made of other tools that may be used to make easier the task of entrepreneurs to raise financial support to their innovation activities.

One of such tools is the identification, dissemination and exchange of good practice. In order to be successful, entrepreneurs need to look at and investigate funding possibilities from different sources. Analyses may be performed in the context of the performance of Venture Capital in Europe; securitisation of guaranteed loans to SMEs; the potential market for IPOs in Europe; support by EU banks to innovation financing and non-regulatory barriers to development of new capital markets. This body of information may be useful to innovation actors looking for funding, by allowing them objectively to identify and pursue funding possibilities which are suitable for their companies.

In addition, experimental pilot projects and networks may be conducted jointly with financiers, covering venture capital models, technology rating and appraisal, spin-out and spin-off from research institutes and universities.

The choice of approach and method is free, but it is a fact that long-term funding can only be secured by establishing closer links with donors and with financing agencies and by their true involvement in innovation projects. One has to be aware of the fact that discontinuity in funding may be fatal to a small company and to its innovation activity. It is therefore important that all efforts should be made to ensure that the process of applying for financial support is fully thought through and that all the different variables are considered.

References

Betz, F. (2003): Managing Technological Innovation: Competitive Advantage from Change. Wiley, Interscience, New York.

Christensen, C. (2003): The Innovator's Dilemma: The Revolutionary National Book That Will Change the Way You Do Business. Harper Business Essentials, New York.

Commission of the European Communities (2004): Innovation in Europe – results for the EU, Iceland and Norway – data 1998-2001. Office of Official Publications of the European Commission, Luxemburg.

Department of Trade and Industry (DTI) (2003): Innovation Report – "Competing in the Global Economy: the Innovation Challenge." DTI, London.

Hussen, D. (1997): The Innovation Challenge. John Wiley and Son, New York.

Leal Filho, W. (ed.) (2002): Prospects of Integration and Development of R&D and the Innovation Potential of Black Sea Economic Cooperation Countries. IOS Press, Amsterdam.

Leal Filho, W. (ed.) (2004): Supporting the Development of R&D and the Innovation Potential of Post-Socialist Countries IOS Press, Amsterdam.

Leal Filho, W. And Ubelis, A.(eds) (2004): Integrative approaches towards sustainability in the Baltic Sea Region. Peter Lang Verlag, Frankfurt.

Thomke, S. (2003): Experimentation Matters: Unlocking the Potential of New Technologies for Innovation. Harvard Business School Press, Cambridge, MA.

Appendix 1

Some Information Sources on Innovation

European

Euro-Abstracts
http://www.cordis.lu/euroabstracts/en/home.html

Innovation and Technology Transfer
http://www.cordis.lu/itt/itt-en/home.html

RTD Info
http://europa.eu.int/comm/research/rtdinfo/index_en.html

CORDIS
http://www.cordis.lu/en/home.html

International

Innovation Magazine
http://www.innovationmagazine.com/

Inventors Digest
http://www.inventorsdigest.com/ME2/default.asp

Innovation Network
http://www.thinksmart.com/

Global Ideas Bank
http://www.globalideasbank.org/site/home/

TRZ Journal
http://www.triz-journal.com/

Chapter 16
Municipal Bonds Market Development in Bulgaria: Challenges and Perspectives

DESISLAVA STOILOVA
Southwest University
Faculty of Economics
2 Krali Marko str.
2700 Blagoevgrad
Bulgaria

Summary

This paper is intended to study the current framework for the development of the municipal bond market in Bulgaria and to suggest solutions for some of the impediments, constraints or lack of regulation that presently exist. Introducing municipal bonds as an attractive and flexible tool for financing municipal budgets and supporting municipal investment projects implies setting up a new institution in Bulgaria, namely the municipal bond market. Despite all challenges and concerns, issuers and investors are preparing to meet at the market. Solving the problems that they face in their entirety and interaction, and facing the challenges will foster the successful development of the municipal bonds market in Bulgaria.

Bulgaria: Broad Perspective

Bulgaria started transition from a centrally planned towards a market economy in 1990. After seven and a half years of a high inflation rate and sharp foreign exchange depreciation the government was forced to establish a Currency Board on July 1, 1997. The very idea of the Currency Board is the tight linking of the local money to an internationally recognized foreign currency. Thus the local money "borrows" creditability from the foreign currency set as a "basis". Bulgarian money (BGL) was tied to the DM at a fixed exchange rate of BGL 1000 per DM. Denomination in 1999 changed this rate to BGL per DM, and since the EURO establishment Bulgarian currency exchange rate has been fixed at BGL 1.95583 per EURO.

The macroeconomic development after the Currency Board's establishment is rather encouraging. The inflation rate dropped to 3-6 percent per year and the foreign currency reserves increased substantially. The Bulgarian economy is in a position to service foreign debt smoothly.

The economic growth issue replaced survival strategy until 1998. Domestic demand was the driving force of real growth in the following years. The stabilization calmed down expected inflation and induced a higher real money demand. However, the supply side factors were planned to be the more important source of growth from 1999 on. Tight fiscal and income policies restrain the aggregate demand and the supply factors gain in importance. The most important supply side factors are structure of ownership and financial markets.

In this context, one of the major problems of the Bulgarian economy is the shortage of financial instruments to channel both domestic and foreign savings toward profitable investment opportunities. The importance of financial market development is pointed out by Joseph Stiglitz in its presentation on The Role of the Financial System in Development: "Researches showed a strong link between economic growth and the depth of the banking system and liquidity of financial markets. The magnitudes of the results are striking: one study found that between 1976 and 1993, countries in the highest quartile of stock market liquidity in the beginning of the period saw GDP grow 3.2 percent annually, compared to 1.8 annual growth for countries in the lowest quartile of stock market liquidity. The difference in GDP growth between countries with high and low financial depth was even larger, 3.2 percent versus 1.4 percent." In a nutshell, the free capital in Bulgaria is far from being automatically allocated to all of the most efficient projects. In effect, bank deposits and governmental securities are the only available financial instruments for households, public and private entities to invest in.

Despite the fact that domestic savings have been estimated at nearly BGL 4 million the National Statistical Institute's estimations have suggested that more than fifty percent of savings were kept in cash and cash equivalents (mostly foreign exchange) instead of investing them in financial assets. The shortage of financial instruments to invest in is a serious problem of the economy. Currently bank deposits and governmental securities are the only available instruments for investment. Even local banks hold a good part of deposits abroad because of the shortage of local instruments to invest in. Bulgaria is in an uneasy position to simultaneously look for a balance of payments support from abroad and to finance the rest of the world.

The Bulgarian capital market is very small and poorly regulated. The daily volume of trade ranges between BGL 20,000 and 50,000. The low level of trade suggests that the capital market does not provide investment opportunities to the net lender of the economy, households. From a broader perspective, municipal debt not only provides an instrument for mobilizing local savings, but also may help boost the capital market development.

Municipalities Needs to Finance Growing Investment Requirements

Municipalities are on the front line to serve the growing needs of the people, from urban infrastructure and transport, to energy, the provision of social services, the upgrading of the environment, and culture. Decentralization raises formidable challenges for the public sector as a whole. First, decentralization must be underpinned by a well-conceived system of fiscal federalism, balancing efficiency and equity considerations through a system of transfers, shared taxes, and local taxes. But fiscal resources alone cannot meet the growing demands placed on municipal governments. Secondly, municipalities face limitations to raise the local tax burden on the private sector and households. They must therefore invent new solutions to meet increasing population needs. Privatizing municipal services and enterprises, entering into concessions agreements, forging partnerships with private corporations to encourage the participation of all population groups into productive economic and social life, tapping domestic and international capital markets to finance growing investment requirements are on the critical path to future development.

Thus the key to meet challenges imposed by the rapid decentralization is to develop budget planning and management capacity, develop the capacity to raise local resources and to access domestic financial markets, creating the framework for private participation in local infrastructure, and promoting local, regional and global partnerships for development. Developing the municipal bond market can be an important step in providing municipalities with an access to the domestic and international capital markets in order to meet the growing investment requirements they face.

In theory, the so-called "golden rule" of the balanced budget prescribes that local authorities should never create a deficit in the budget for the purpose of covering current costs. But the same rule allows and in some formulations even promotes prudent borrowing for capital purposes. So, why can local governments borrow to finance their investments, and why should they?

The most important arguments in favour of local government borrowing are summarized by Swianiewicz (2003) as follows:

- Inter-temporal, inter-generational and geographical equity. The costs of a project undertaken by a local government are incurred as soon as the project is implemented, but the benefits derived from the investment are spread out over a longer period. If capital projects are financed out of current revenues, then some local taxpayers who helped to finance the project through their local taxes may not benefit from them in the future (owing to migration out of the city, for example). At the same time, those who benefit from the project may not have participated in financing it if they moved to the city after it was completed, or if it was completed either when they were small children or before they were born. By financing such projects through bond issues or bank credit, local governments can ensure that most users pay for the benefits either through local taxation or directly in the form of user charges. In this way, payments from current users are partially used to repay the loan.
- A closer relationship between those who benefit from the project and those who pay for its completion supports optimal allocation of resources.
- Benefits from the acceleration of local development due to the borrowed funds quite often overshadow the cost of debt servicing. It may happen that benefits from an appropriate project implementation, such as higher price or rents, wider scope of interested investors, quicker economic development resulting in multiple effects that attract new projects, providing additional jobs and tax revenues, are much greater than the cost resulting from interest payments to the investors or banks.
- Reduction of operational costs. If local infrastructure is too old it will be more profitable to borrow funds and build new and more reliable one instead of repairing it. Benefits of this solution are higher reliability, savings in the costs of repair and the employment of service staff and other secondary advantages. In fact, costs related to borrowing may even be less than costs involved in maintaining the old infrastructure for a longer time.
- Access to grants from European and other development funds. This is one more rationale for borrowing, specific to Central and Eastern European countries. There are several investment grants available for local authorities, but a necessary condition is to provide matching funds that usually must equal at least 25% of the total project costs (as in the case of SAPARD or ISPA projects). Borrowing may be a means of increasing local capacity to apply for these development grants.

However, there are also potential hazards related to local government borrowing, of both a microeconomic and a macroeconomic nature. The microeconomic danger is related to the potential for excessive indebtedness of some local governments, leading to serious difficulties in the repayment of loans and possibly jeopardizing the provision of vital public services. The macroeconomic limitation is related to the fact that local governments contribute to the overall level of public debt, which in turn may have a negative effect on inflation and other important parameters of the national economy.

Why should local governments not borrow to cover their operating expenditure? There is a common agreement that borrowing in order to cover current expenditures is acceptable only in very rare, specific cases – usually for very short periods, to cover deficits arising from uneven cash flows within a budgetary year. The most typical arguments for the importance of maintaining a balanced operating budget can be summarized as follows:

- Borrowing on operating spending would lead to an unmanageable debt burden. It would quickly lead to the rolling of loans (using new loans to finance the payment of an earlier debt's service) and to a very serious problem of excessive indebtedness.
- Using current revenues to cover current costs prevents the local public sector from growing beyond its optimal size, which may be defined as the fiscal burden that taxpayers agree to bear in order to finance the desired provision of public goods. Borrowing creates a short term fiscal illusion, in which the demand for public services is artificially high because it is distorted by the supply, financed not by local tax effort but in part by credits, bonds, etc.
- An unbalanced current budget may result in negative macroeconomic and microeconomic consequences, since private investments may be crowded out. It is clear that public sector borrowings draw on the pool of limited financial resources available. Local government issues are more attractive to investors than private issues, because giving credits to public entities implies lower risk. This competition for borrowing from the public sector exerts an upward pressure on the interest rate, making private investments more costly. Increasing budget deficits negatively affect expectations on inflation, which add more to the upward trend in interest rates. Of course it is worth noting that when local governments offer loans in order to finance investments, this negative "crowding-out" consequence does not occur, unless someone postulates that public investments are less productive then private ones.

This brief overview brings us to the conclusion that in general both investors and issuers (municipalities) are interested in entering the municipal bond play. Whether and when that play can take place will be determined by the scene (macroeconomic and regulatory framework) and by the interaction among players, both analyzed below.

Potential Issuers: Addressing the Problems of Municipal Budgets

Bulgaria has undergone a process of gradual public sector decentralization during the last thirteen years. The process started in 1991 when the Local Self-Government and Local Administration Act was adopted. According to the law, our country has two tiers of sub-national administrative government – regions and municipalities. The region is an administrative unit of the Central Government. It comprises several neighbouring municipalities. The region can be created and liquidated only by law. A regional governor, who is appointed by the Council of Ministers (the Government), governs the region. Presently 28 regions exist, which were established by splitting the 9 regions, which existed before 1999. The Regional Administration is a decentralized state authority on regional level. It has three main functions: to manage the state properties on the territory of the region, to monitor compliance of the municipal council decisions with law, and to implement the state policy on local level. In addition to their controlling function, which was intended mainly to combat and curtail corruption, the Regional Administrations were formed to foster development and to unite municipalities to work together on large-scale projects and common causes. Although the fact that the regions coordinate national and local interests they do not perform executive functions. So the Regional Administrations are mediators between the central and local authorities. They do not have independent budgets and financial independence, and do not provide public services to the population. The budget of the regions is determined by the Central Government and the funds are used for the operation of these units. The Regional Administrations do not have their own revenues. Their budget is entirely formed by subsidies from the central budget. The budget of the Regional Administrations is 0.1% from the consolidated public expenditures and 0.04% from the general domestic product (GDP).[1]

According to the Constitution the municipality is the main (and by now the only one) tier of local government in the country. The municipality is a legal entity. It has its own budget and property, which can be used for serving its interests. The bodies of local government – the Municipal Council and Mayor – are elected directly by the population of the whole municipality. Despite the fact that Bulgaria is a small country (approximately 111,000 sq. km), presently there are 263 municipalities. Although legal possibilities for splitting and merging of municipalities exist, their number is not very dynamic. The characteristic features of the municipalities are as follows: the average territory is approximately 422 sq. km, the average population is approximately 30 000 people and the average number of settlements is approximately 20.[2]

In fact, the gradual decentralization process results in increased local self-governance and ever-increasing responsibilities for the municipalities to provide solutions as to how to meet the population demands (ranging from transportation to cultural events and leisure). At the same time, data supplied by the National Association of Municipalities in the Republic of Bulgaria (NAMBR) shows that local budgets relative share to the GNP has decreased to 6% in 2003 as compared to 12.3% in 1991 and 8.9% in 1999.

Table 1: Bulgaria: Macroeconomic and Budgetary Indicators

Indices	**1999**	**2000**	**2001**	**2002**
GDP (BGL million)	22776.0	26752.8	29617.7	32323.7
Consolidated State Budget (BGL million)	9912.2	11334.3	12096.5	12732.5
Municipal Budgets (BGL million)	2022.3	2178.1	2113.9	2388.5
Consolidated State Budget / GDP (%)	43.52%	42.37%	40.84%	39.39%
Municipal Budgets / GDP (%)	8.88%	8.14%	7.14%	7.39%
Municipal Budgets / Consolidated State Budget (%)	20.40%	19.22%	17.47%	18.76%

Source: National Association of Municipalities in the Republic of Bulgaria: Financial Database (2003). In http://www.namrb.org/findb/fin01.htm

To a large extent, this is due to the fact that local authorities have been assigned various responsibilities, while municipal revenues coming from state transfers have not been particularly increased, rather, these tend to decrease in real terms. By virtue of the current law, municipalities are fairly restricted to exercising an influence on their revenues coming from local taxes and fees. It is, therefore, imperative that municipalities continue looking for new sources, to increase their revenues and to raise funds in order adequately to meet their increased responsibilities. So far privatizations of municipal properties and/or streamlining of its management and maintenance as well as extending concessions have been utilized as major methods to increase revenues. For various reasons municipalities do not have exten-

[1] Club "Economics 2000" (2002), Needs Assessment of Sub-national Governments Draft Report, Sofia, p. 10.

[2] Ibid p. 7.

sive experience in raising funds as yet. It could be maintained that the amount of bank loans extended to municipalities has been nominal so far. Therefore, it could be expected that other, less traditional approaches, such as equity participation and issuing municipal bonds would grow in importance in the future.

Although the local public sector provides basic public services to the population, municipalities have very limited powers for setting local revenues. In order to facilitate further analysis municipal revenue assignment according to the Municipal Budgets Act is presented in *Chart 1* below.

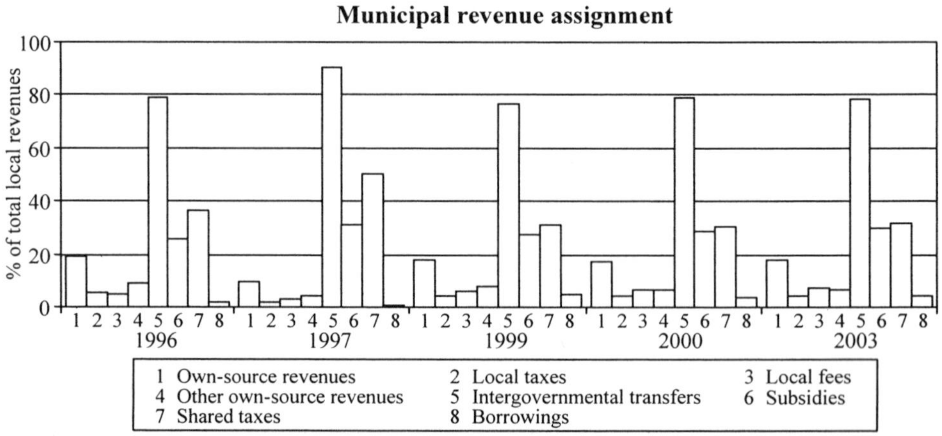

Fig.1: Municipal revenue assignment

Source: Club Economics 2000. (2002) Needs Assessment of Sub-national Governments Draft Report, Sofia, p.11 and Ministry of Finance Database (2003). In www.minfin.government.bg/

Bulgarian municipalities are fiscally weak. Because of overly restrictive interpretations of the Constitution, local governments are not allowed to impose their own taxes or levy surcharges on central government's taxes. For this reason, Bulgarian municipalities do not have a lot of opportunities to show flexibility with regard to revenues. Revenue analysis of municipal budgets clearly indicates that since 1991, state transfers have been of crucial importance. Virtually, Bulgarian municipalities rely heavily on the intergovernmental transfer system. The revenues from state transfers comprise about 80% (namely, 94% in 1991, 78% in 1994, 90% in 1997, 80% in 2000, and 78% in 2003) of the total budget revenues of Bulgarian municipalities. As a comparison, the average ratio between the two main local revenue sources across all Western European countries is 37% (transfers) to 50% (local own-source revenues).[3] According to the economic and financial point of view, such a proportion allows local governments to be compensated for the vertical imbalance, horizontal imbalance and spillover effects. But according to the political and institutional point of view, the dominant role of the transfers gives the central government significant political and financial control over local governments.

The intergovernmental transfers in Bulgaria are not competitive. They are allocated as entitlement. As financial flows, they descend from two sources: shared taxes and state subsidies. The revenues from the shared taxes amount to as much as 51% to 30% of the total municipal budget revenues, while the percentage of the state subsidies to the total munici-

[3] J. McCullough, T. Spofford, E. Savov, S. Ivanov (2000), Bulgaria: Comprehensive Municipal Finance and Fiscal Reform Proposal: Final Report, Sofia, p. 6-7

pal budget revenues has slightly increased from 22% in 1991 to 30-31% in 2003. State subsidies are allocated based on a formula, which is stated in the Annual State Budget Acts and since its introduction in 1993 it has been changed each year. Basically, the formula takes into account the expenditure needs and revenue capacity of local governments. Traditionally, Bulgarian municipalities are granted general supplemental subsidy, general equalization subsidy, and target capital investment subsidy.

The most important shared tax revenue is the personal income tax. It was shared among the state and municipalities in 50:50 ratios during the previous years. Since the beginning of 2003 the personal income tax has been defined as municipal revenue entirely, except for 35 municipalities with great fiscal capacity. They receive a certain percentage of the personal income tax revenues, which varies widely from 17.6% to 99.1% in any particular case.[4] In spite of this attempt at equalization, significant interregional disparities are inescapable, because the individual income tax is a progressive tax, collected by withholding at source and the tax bases are unevenly distributed, favouring the richer local governments. From the point of view of the grant taxonomy the shared personal income tax in Bulgaria is an unconditional, open-ended grant.

Revenues from the company profit tax were also shared during the previous years, but since the beginning of 2003 the corporate income taxes have been defined as state revenues entirely. Municipalities do not receive any share of excises, customs duties or VAT, which are the main central budget's revenue sources.

Where own revenues are concerned, municipalities are only allowed to determine the amount of local fees[5] within certain margins and to streamline the process of municipal property management (including privatization, concessions, rents, sales of real estate and the like). From the figures above it can be concluded that the share of own-source municipal revenues in the total municipal budget revenues in Bulgaria is only about 17-18%. In this context, it must be pointed out that Bulgarian local governments do not have power to determine the type, the rates or the basis of the local taxes. In fact, the only reason they are called "local" is that they are stated by the Local Taxes and Fees Act. In terms of modern public finance, if local governments do not have any say in the "design" of the local tax, it could not be considered as a local tax, but as a special transfer. Thus, according to cited taxonomy, up to 90% of the Bulgarian municipal budgets is formed by transfers and is determined directly by the central government.

Neither do municipalities have a lot of opportunities to show flexibility with regard to expenditures. Over 59% of their funds are directed to financing health care, education and social care activities. These three are a matter of shared competence between municipalities and the state. In practice municipalities provide the necessary financing, while the state (through the respective authorities) determines the amount of the average salary, the number of persons employed and appoints the management of municipal bodies (although the mayor or municipal council is entitled to suggest the persons appropriate for the respective position). While analyzing municipal expenditures, it has been established that 2/3 of total municipal funds are used to finance development activities in the above-mentioned spheres, namely, health, social care and education. It is interesting to note that the municipalities have financed up to 70% the total amount of funds used to sustain education in Bulgaria in

---

[4] They are listed in The Annual State Budget Act of the Republic of Bulgaria for Fiscal Year 2003, published by the State Official Journal, Issue 120, December 29 2002, Appendix 3

[5] Notwithstanding that local fees in practice are payments for rendered services, municipalities are entitled to define the amount of local fees through a decision taken by the Municipal Council within minimum range, regulated by normative act.

the 90s. In recent years this share dropped to 55-57%. The change is due to the fact that expenditure on the secondary professional schools and the schools for disabled children started to be financed from the central budget. A similar trend is monitored in the field of health care, but the reason here is the gradual introduction of the health insurance system, which started in 1999.

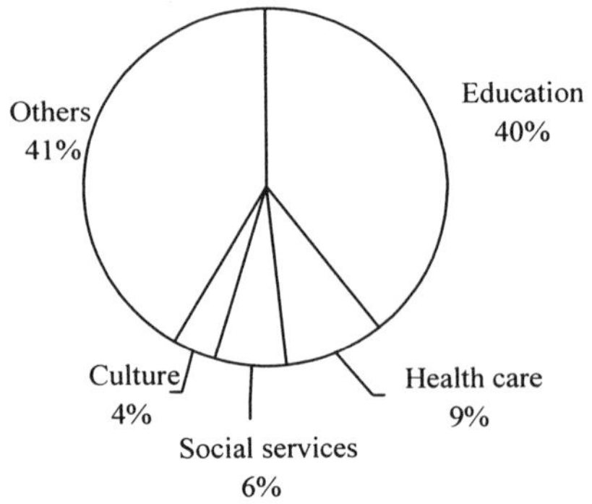

Fig. 2: Municipal expenditures assignment

Source: National Association of Municipalities in the Republic of Bulgaria: Financial Database (2003). In http://www.namrb.org/findb/fin04b.htm

The share of raised funds (including loans from financial institutions, issue of municipal bonds, interest-free loans from the state budget, loans from municipalities, and off-budget funds) is fairly insignificant. For instance, their share for the last 5 years varies between 5 to 0.3% of the total municipal revenues. Loans taken from financial institutions have been nominal, but for the case of the Sofia municipality. In practice 81% from the borrowed funds are used by Sofia (the capital city), and 17% by the largest municipalities.[6]

Bulgarian municipalities to a great extent rely on obtaining financing from international organizations to support infrastructure projects. The latter, as in the case of Poland, might delay the municipal bond market development since easy access to low cost external financing may reduce the municipalities' motivation to show more initiative. Furthermore, the Ministry of Finance (MoF) following the Annual State Budget Act usually approves and/or elaborates municipal investment programs and its financing. As a rule, target subsidies from the state budget are used to finance investment projects. A good example in this respect is the Kyustendil Municipality. A PHARE program covered all schools on the territory of the municipality, the water supply system renovation was financed through the resources of a loan extended by the World Bank, referred to as the water loan. Presently, these sources of revenues are about to be exhausted. The existing four or five municipal companies are no source of revenues, all of them functioning below the break-even point.

Municipal finance managers clearly prefer to finance budget deficits with the cheapest resources available, and these are ranked in the following order:

- Revenues coming from municipal off-budget accounts (interest not due);
- Obtaining loans from other municipalities (no interest is due);
- Borrowing from banks.

In fact, debt is rolled over. At the end of fiscal year, debt is paid back to the off-budget account. Obtaining loans from other municipalities is the second best alternative. Bank loans

---

6 Club „Economics 2000" (2002), Needs Assessment of Sub-national Governments Draft Report, Sofia, p.11

are much more costly and most municipal finance managers believe that the cost of bond issue would be even higher. Currently, this financial instrument is new to them and they feel more inclined to rely on borrowings from other organizations and even bank loans than bond issuance. It must also be noted that all officials who considered issuance of municipal bonds as a means of fund-raising have stated that the choice between obtaining a bank loan and issuing bonds would depend on the type of investment project and should be made on a case by case basis. Given the fact that Bulgarian banks extend investment loans for up to two years and the latter are very short of supply, no wonder that the municipal officials of Svishtov and Troyan (two relatively small towns) have chosen to launch public issuance of general bonds to finance renovation of the local water supply systems regardless of the fact that the transaction cost of such issue could be higher as opposed to obtaining a bank loan.

In addition, employment of new financial instruments, namely municipal bonds, requires better quality of management at the local level and changes dramatically the relationships between the municipalities and central government. Undoubtedly, the financial instruments issue is a final point of many other activities, which strengthen the principles of local self-government and regulate the relationship among the municipal bonds market participants: municipalities, business, and investors. The transparency of local government and the public debate are very important for defining the local priorities and the public support of the municipal investment program. For example in 1999 the municipality of Sofia issued Eurobonds on the Luxembourg stock exchange with a nominal value 50 million euros. The debt was placed successfully with foreign investors within a couple of days and was paid back without any problems. However, the point is that the population of Sofia knows almost nothing about the investment projects and financial plans of the municipality.

In conclusion, as proven by the recent examples of the municipalities of Sofia, Svishtov, Troyan, Varna, Dupnitza, Bulgarian municipalities are able to issue bonds and this financial instrument has potential. In order to facilitate the development of the municipal bond market the major problems of the municipal budgets have to be addressed. The basic problems can be summarized in the following groups:

- Insufficient authority to manage municipal own revenues and strong dependence on state transfers;
- Limited potentialities to manage and exercise an influence on expenditures in fields of shared competence, where state authorities are entitled to determine levels of salaries, number of personnel, management of municipal budget institutions, etc., while the respective municipalities provide the financing;
- Lack of sufficient transparency and constantly changing mechanisms employed in distribution of subsidies among municipalities that restrict municipal potentialities to make independent, stable and long-term budget projections;
- Lack of sufficient knowledge, skills and experience to practice financial management, or efficiently mobilize and utilize financial resources, and apply alternative sources to fund local services.

Reviewing the national policy in the field of municipal finance could potentially solve the first three problems. Amendments in the Municipal Budgets Act are needed to stabilize the interrelationship between local and central budgets. The last problem requires constant improvement of knowledge in the field of public finance that can be achieved through various types of technical assistance or support by the National Association of Municipalities in the Republic of Bulgaria (NAMBR) or other NGOs.

Ability of Bulgarian Municipalities to Receive Loans and to Service Debt Payments

From a broad perspective the capability of municipalities to repay their debts is dependent on their tax revenue base. Municipal tax capacity determines both the amounts of their own revenues and the revenues coming from shared taxes. It is crucial for the municipality that local companies headquartered within the municipal territory should be operative and profitable. However, taxes are collected by the tax administration, which is directly subordinated to the Ministry of Finance, and municipalities have limited capacities to prevent tax evasion and to increase tax efforts. In the long run, the status of municipal budgets is dependent on both the macroeconomic situation in the country and the capacities of local population and entrepreneurs to successfully meet ever-occurring challenges. Given the positive outlook for Bulgaria it can be expected that the capabilities of municipalities to repay their debts should increase in the future.

From a narrower perspective the basic criterion for municipal eligibility to receive loans and to issue bonds is its creditworthiness. The potential of a municipality to generate own funds is, on the one hand, a basic factor applied in determining its creditworthiness; on the other hand, owing to restrictions, defined in the Municipal Budgets Act, municipal potential for borrowing is limited: the annual interest and principal debt payments cannot exceed 25% of its own revenues plus the total equalizing subsidy granted to the municipality. Undoubtedly, this regulation puts larger municipalities in a better position to raise funds.

Is regulation of local government debt really necessary? It may be argued that it is enough to rely on financial market discipline. This line of argument suggests that the adoption of legal rules might be redundant, since tighter bond market conditions – in particular, higher interest rates – already impose effective sanctions. If the total debt of local governments in the country grows too high, this will indeed be the case. Also, investors will refuse to buy municipal bonds or will demand a very high interest premium for those municipalities that borrow more than they can carry out effectively.

Dafflon (2002) suggests, however, that in practice there are several doubtful assumptions behind the arguments presented in the previous paragraph. One such assumption is related to the adequacy of the information possessed by investors. Another practical observation is that investors usually assume that no real risk exists in case of local government default. They believe that the state will eventually pay the debt. Although a local government bankruptcy is technically possible in several countries, it is rarely observed in practice because it is so politically unacceptable. For these reasons, then, external regulations and control of local borrowing may play a positive role supporting the local credit market.

The recent regulations of the Maastricht stabilization pact related to the introduction of a "Euro zone" have brought a new element into discussions on local indebtedness. The Maastricht agreement limits the overall level of public debt to a maximum of 60% of GDP, and also limits the annual total public budget deficit to 3% of GDP. It should be noted also that the level of public debt is usually much higher at the central than at the local level.

At present, the total debt of all Bulgarian municipalities does not exceed BGL 70 million[7] or up to 1% of the consolidated state budget and 0.2% of GDP. It can be concluded that the low share of the own municipal revenues is one of the most significant impediments for a further development of municipal bonds market in Bulgaria. One way to overcome such an

---

7 Joint Research/Exchange Program (1999) A Set of Recommendations on the Practical Arrangements for the Issue of Municipal Bonds in Bulgaria Working Paper, JREX, Sofia

impediment is the uninterrupted application of measures to improve local self-government, adequate to the municipality's financial interests and potentialities. New solutions for increasing own municipal revenues should be sought, for example municipalities should have more freedom and discretion in determining the amount of local taxes and fees, especially when those arise from user charges or rendered services.

Larger municipalities are clearly in a better position to receive loans and to issue bonds. They are in a good bargaining position in receiving state transfers. The probability to have larger target capital investment subsidy from state budget corresponds with the size of the municipality rather than with the needs for investments in the local infrastructure and services. It can be assumed that at this stage only a very limited number of financially independent municipalities can afford to invest in infrastructure projects. Therefore, it can be expected that only a very restricted numbers of all 263 municipalities in Bulgaria, will be able successfully to issue bonds under the current circumstances, since they will have to simultaneously meet the following requirements: to be relatively large and to be financially independent.

Potential Investors

Banks, insurance companies, pension funds, private companies and households constitute the list of potential investors in municipal bonds. The legal framework on municipal bonds does not adequately protect the interests of investors, especially in the event of default. There are two main reasons to explain this: first of all, the event of default is not adequately defined, and second, common practice of bank foreclosures in cases of defaulting loans clearly indicates that legal protection provided to lenders is extremely inefficient. Owing to the reasons mentioned above, it is crucial to investors that they are enabled to monitor that funds raised by bond issue are efficiently invested and have the measures to restrict risks of defaulting. The best way to achieve this (apart from executing a perfect bond agreement) is to have access to prior experience and information on municipal activities.

At present, banks are the only institutions that are in position to invest abroad. The net foreign assets account for more than 40% of the consolidated balance sheet of the banking system (excluding the Bulgarian National Bank). Supposedly, this is due to the lack of domestic financial instruments and the artificially maintained low primary interest rate. Municipal bonds are new instruments for the Bulgarian financial market and in order to attract bank investment interest bank personnel would need to become better acquainted with municipal finance. In Poland, for example, bank personnel were literally induced to undergo training in the field of municipal finance as well as bond issuance risks. It must be noted that the major driving force to develop municipal bond market was competition among banks, especially competition between local and foreign banks. None of these is in place in Bulgaria now, but other countries' experience may differ.

The few banks that have been allowed to service municipal budgets have access to much more information with regard to municipal activities, as opposed to those that are not allowed. In addition, banks are better suited to monitor execution of investment projects as compared with a multitude of individual investors. That is why it is very probable that banks will be the major investors in municipal bonds in the immediate future.

It can be expected that in the long-term perspective institutional investors like insurance companies and pension funds will most certainly show a growing interest in investing in bonds as these are the only investors of long-term financial resources. Currently, portfolios of insurance companies and pension funds consist basically of state securities and bank deposits, although they are allowed to invest their resources in low-risk financial instruments.

At present, it cannot be expected that business companies will show exceptional interest in investing in municipal bonds unless serious tax preferences are implemented. Not to mention the fact that companies themselves suffer severe shortage of credit resources and working capital. Current analysis reveals that lately intercorporate indebtedness has reached extremely high levels.

Households and individuals have been ruled out as investors in municipal bonds in recent years, owing to the assumption that people has not as yet overcome their mistrust in the financial system, a mistrust originating from the high inflation levels, the existence of financial pyramids and the severe bank crisis of 1996-1997. However, the low interest rates of bank deposits and state securities can be a precondition for channelling households' savings toward more profitable investment opportunities.

In conclusion, it can be summarized that a key factor for the promotion of the purchasing of municipal bonds, apart from availability and execution of a sound contract, is the availability of sufficient information and transparency of municipal finance, currently unavailable. The problem can be overcome by the implementation and strict observation of legally defined requirements that will ensure the transparency of public finance as a whole, as well as the launching of programmes for providing information to potential investors about municipal activities, preparing and presenting municipal financial statements. The Ministry of Finance experience in launching state securities of 5-year maturity proves that the local market is not ready yet to purchase long-term bond issuance of fixed profitability which implies that in order to sell better, municipal bonds should have a variable interest rate. The most experienced investors currently are Bulgarian banks, a fact that comes to support the statement that the most favourable (and the cheapest) way to develop municipal bond market would be private placements.

Because the regulatory framework concerning municipal bonds issue is more inefficient and scarce as compared to corporate bonds regulations, the separate act on municipal bonds should be elaborated. Another unfavorable factor is that municipalities are entitled to issue general bonds only. They should be allowed to issue revenue bonds as well. Because revenue bonds are project-oriented, they would be favourable for the smaller municipalities, since larger municipalities are in a better position to raise funds. Furthermore, it is doubtful whether private placements are permitted under the present legislation. This can impede the development of the municipal bond market since private placements involve lower transaction costs.

Conclusion

Building a better legal and institutional framework for fiscal decentralization is critical to advancing a wide array of objectives aimed at achieving more effective and responsive local government in Bulgaria. From the narrower perspective of municipal bond market development, clearly progress on this front is a prerequisite for municipal bonds to emerge as a predominant means for financing local investment needs. The recommended reforms should be designed to:

- Strengthen municipal authority and capacity to increase revenues and control operating expenditures, with an eye to generating a meaningful level of annual net savings in the operating budgets on a predictable and recurring basis;
- Motivate and reward those municipalities that take the initiative to borrow and invest, particularly for capital improvements that pay for themselves through savings in operating costs.

Specific reforms which would advance the above objectives include:

- Establishing a budget with separate operating and capital accounts, and multi-year capital investment plans, that would include the ability to carry forward surpluses;
- Granting greater local discretion to set local taxes and fees;
- Introducing a more transparent and predictable system of transfers; and
- Allowing more local responsibility for the forecast of own source revenues.

From the investors' point of view some additional recommendations should be made:

- To simplify due diligence for lenders. Financial institutions will be more inclined to take the trouble to underwrite municipal credits if they have ready access to reliable information on municipal financial status – particularly in respect to accrued and contingent liabilities and outstanding indebtedness;
- To give investors confidence in the financial procedures, systems, and management capacity of local government that will help ensure timely repayment.

In conclusion, Bulgaria has a unique and timely opportunity to establish a well-conceived policy and legal framework in order to advance the municipal bond market development. Despite all challenges and concerns, issuers and investors are preparing to meet at the market. Solving the problems that they face in their entirety and interaction, and facing these challenges, will promote the successful development of the municipal bonds market in Bulgaria.

References

Club "Economics 2000" (2002), Needs Assessment of Sub-national Governments Draft Report, Sofia

Dafflon, B. (2002) Local Public Finance in Europe: Balancing the Budget and Controlling Debt. Studies in Fiscal Federalism and State-Local Finance Series. Cheltenham-Northampton: Edward Elgar.

Davey, K. (2002) Balancing National and Local Responsibilities: Education Management and Finance in Four Central European Countries, LGI, Budapest

Epstein, P., G.Paterson, J.Pigey, S.Sherer (2000) Municipal Credit Market Development in Bulgaria: Policy and Legal Framework, TUI, Sofia

McCullough, J., T. Spofford, E. Savov, S. Ivanov (2000) Bulgaria: Comprehensive Municipal Finance and Fiscal Reform Proposal: Final Report, LGI, Sofia

Swianiewicz, P., (2003) Foundations of Fiscal Decentralization: Benchmarking Guide for Countries in Transition, Local Government and Public Service Reform Initiative/Open Society Institute, Budapest Joint Research/Exchange Program (1999) A Set of Recommendations on the Practical Arrangements for the Issue of Municipal Bonds in Bulgaria Working Paper, JREX, Sofia

Chapter 17
Innovation Tomorrow: Excerpts of the Official EU Policy Document on Innovation

WALTER LEAL FILHO
TuTech Innovation
Harburger Schlossstrasse 6-10
21079 Hamburg
Germany

Summary

This final chapter presents excerpts of the executive summary of the official EU policy document "Innovation Tomorrow"[1]. The contents will be of special interest to those involved in innovation policy and support in all European countries.

Introduction

The Lisbon Summit in 2000 set a critical strategic goal for the European Union. The European Union should, by 2010, "become the most competitive and dynamic knowledge-based economy in the world, capable of sustainable economic growth with more and better jobs and greater social cohesion".

How is Europe to achieve this goal? This question inspired the report. It focused specifically on the topic of innovation and its relation to policy and policy-making.

Innovation is influenced in significant ways by areas of policy over and above "innovation policy", as it has been heretofore understood. The argument is that a "third generation innovation policy", which recognises the centrality of innovation to effectively all policy areas, must be developed.

Innovation is a central element of the Lisbon objective. Innovation is required to become, and to remain, "competitive and dynamic". It is now well known that innovation is a central element of economic performance. Its growing importance makes it a core feature of the knowledge-based economy. Innovation has also facilitated the development of the knowledge-based economy. (For instance, enabling the shift to service-based activities and widespread use of new Information and Communications Technologies.) But the nature of innovation is also changing in the knowledge-based economy.

---

1 "Innovation tomorrow – Innovation policy and the regulatory framework: Making innovation an integral part of the broader structural agenda", Louis Lengrand & Associés, PREST (University of Manchester), ANRT – France, Innovation papers No 28, Directorate-General for Enterprise, EUR 17052.

Knowledge-based activities stimulate new kinds of innovation, and also allow for innovation processes to be reconfigured. Many other influences are also reshaping innovation and the boundaries between various areas of policy. We cannot assume that established policies for encouraging innovation are adequate to the new conditions. Nor can we assume that the ways in which other policy areas have taken innovation into account are still adequate.

Innovation in a knowledge-based economy is diverse and pervasive. It is not just based on research, or science and technology, or enterprise and ingenuity – although all of these remain very important contributing factors. Innovation – especially successful innovation! – also depends on organisational, social, economic, marketing and other knowledge. It frequently requires intellectual and artistic creativity. There is an increasing emphasis on such "intangible assets" within firms. Their role in allowing for regions and nations to become dynamic participants in the knowledge-based society is also acknowledged. But problems in properly valuing such assets have become strikingly apparent in the financial turbulence of recent years.

Regulatory and institutional reform is being – and should be – encouraged across a wide range of policy areas. In general this has not been undertaken with ideas of promoting innovation in mind. The rationale for reform has been that outdated, over complex, or over-prescriptive regulations are encumbering business. The assumption has been that reforms will more or less automatically foster innovation by freeing up the resources consumed in this way. However, not all regulations do simply inhibit innovation – some shape it, some actually foster it. Furthermore, innovation can also be actively encouraged by innovation-friendly reforms in various policy areas. This has been recognised in some fields and specific cases, but it is a more general state of affairs. Furthermore, lessons can be drawn from experiences in different countries and different policy areas. There may well be synergies between and across policy areas, too, so that co-ordinated reforms could have more of an impact than one-off instances.

Similar implications to those concerning regulatory reform arise from efforts to reform governance processes. Here, the relations to innovation especially concern the legitimacy of research efforts, and the public attitudes to expertise, innovations and innovation related policies. There are potential tensions between efficient and flexible innovation policymaking, and the need for openness, transparency and dialogue.

The "third generation innovation policy" will have to confront the significance of a wide range of policies for innovation in the knowledge-based economy, and establish ways of building analysis and action relevant to innovation into all of these policy areas. Policy-makers and other stakeholders concerned with innovation, and other topics that impinge upon innovation, have to deal with numerous objectives and immediate pressures.

The challenge for the third generation innovation policy will be to maintain a core emphasis on innovation across the board. This emphasis will need to be sensitive to the continuing transformation of innovation processes in the knowledge-based economy. It cannot be a matter of a one-off set of reforms – attention will need to be paid to ongoing change in the nature of innovation. This sensitivity will have to be maintained while innovation is being built as a criterion into a series of policy areas.

Innovation performance, innovation policy, and the knowledge-based economy

The innovation performance needed by Europe requires us to constantly enhance sustainable capacities to sustain productivity growth. The report document brings together ideas and conclusions emerging from various lines of study. In order to put the analyses and case studies into context, the report examines the main characteristics of a knowledge-based economy. Three broad groups of features of the knowledge-based economy (along with associated features such as globalisation) have a considerable bearing on the nature of innovation:

- The rise of services and intangibles
- The ongoing rapid development of Information and Communication Technologies and the Information Society
- The new roles of knowledge, organizational learning and human resources.
- Innovation tomorrow

The current European Innovation Policy, based on the Innovation Action Plan arising from the December 1995 Green Paper on Innovation, included, among its three principal objectives, the goal of establishing "a legal, regulatory and financial framework conducive to innovation". Innovation is a contributing element of the knowledge-based economy. It is also something that is transformed by the new practices and organisations that are emerging here. Regulatory and institutional reform is also related to the rise of the knowledge-based economy. It reflects an effort to bring knowledge into policy design and implementation, so as to make policies more effective (and, indeed, to find means of achieving policy objectives that are more effective than conventional regulations, where appropriate).

There are several reasons for such regulatory reform:

1. improving internal efficiency;
2. improving service delivery,
3. simplifying procedures, so as to reduce confusion (e.g. about grants and incentives), and the administrative burdens (e.g. in SMEs and other users).

Regulatory reform can achieve increased effectiveness and thereby help to solve challenging economic and social problems. One way in which it does so is by combining actions and strategies in related policy areas. Similar effects have been achieved in businesses, by removing the boundaries of functional departments and focussing instead on business processes.

Current innovation policy – "second generation innovation policy" – emphasises the importance of the systems and infrastructures that support innovation. These are influenced by many policy areas, in particular research, education, taxation, IPR and competition policy. But these policy areas are not exclusively focused on innovation matters; they are not the only ones to impinge on innovation; and the need to work together is not always recognised.

The Target: third generation innovation policy

The 1st generation of innovation policy was based on the idea of a linear process for the development of innovations. This process begins with laboratory science and moves through successive stages till the new knowledge is built into commercial applications that diffuse in the economic system. The emphasis of policy was on fostering critical directions of scientific and technological advance, and enhancing the flow of knowledge down along the innovation chain.

Second generation policy recognises the complexity of the innovation system, with many feedback loops between the different "stages" of the process as outlined in the 1st generation model. It also gives more recognition to the generation and diffusion of innovations within what have become known as "innovation systems" (national, regional, sectoral, etc.). Policy seeks to enhance two-way communication across different points in the innovation "chain", and to improve innovation systems in ways that can better inform decisions about research, commercialisation, technology adoption and implementation, etc.

Even though 2nd generation policies still have to be embedded in many agencies, the contours of a new generation of innovation policy are now becoming apparent. Such a new generation of policy would emphasise the benefits of co-ordinating actions in policy areas, and making innovation – and innovation-friendly policies – one of the core principles of this.

Thus "third generation innovation policy" would place innovation at the heart of each policy area. The report outlines the case for doing this, in each of a set of policy areas considered. There are of course many differences of detail from one policy area to another.

The common aim is to maximise the chances that regulatory reform will support innovation objectives, rather than run the risk of impeding or undermining them. But to accomplish this means fusing two sorts of knowledge. First is knowledge about (the changing nature of) innovation processes and innovation policy. Second is knowledge about (the rationale and reform processes underway in) the specific policy areas. This fusion should be assisted by the sort of broad-brush review presented in this report. Systematic research on the topics discussed here is, however, relatively sparse. Only a few areas, such as competition policy and IPR policy, have attracted much attention in terms of their relations to innovation – and such attention has typically been directed at very few of the links that exist. This limits the depth with which the topics can be reviewed. Nevertheless, a number of themes do clearly emerge suggesting how the knowledge-based economy jointly challenges innovation and other policy areas. But accomplishing the objective of fusion requires considerable depth of understanding about specific policy design and implementation issues.

This can only come from extensive in-depth involvement with the specific areas. Efforts to specify what is needed from the basis of one or other academic discipline – be it economics, political science, or even innovation policy studies – are bound to be too narrow. Furthermore, such efforts are unlikely to be politically persuasive to those actually encharged with policy design and implementation.

Thus, while the broad contours of a "third generation innovation policy" can be outlined, a serious effort to articulate and fuse the bodies of knowledge that need to be brought together to add depth to this is still required. Indeed, such a task of articulation and fusion is a feature of the innovation process itself. In the innovation process, new ways of collaborating to produce innovations are pursued by different actors, none of whom possess sufficient range of in-depth experience to put together all the pieces required to make successful innovations. Similar challenges confront policy innovations.

The new policy cannot be introduced quickly and simply by announcing that a third generation policy is overdue. There will be a need for leadership, education, examples, guidance and co-ordination services. Responsibilities for bringing together relevant evidence and benchmarking progress, for ensuring that all policy areas do take on the message of the centrality of innovation, and learning the lessons of experience and experimentation need to be established.

The creation of a third generation innovation policy should be a political objective. The steps described above should be accompanied by a regular review of progress in this direction, perhaps on a biannual basis. This should be the opportunity for widespread debate on the critical principles and issues arising. Such knowledge-based policy is essential if the knowledge-based economy is to be achieved in line with the Lisbon objectives.

Main Policy Recommendations

A clearer understanding of the knowledge-based economy and of the importance of innovation and innovativeness within it, must continue to be developed, and should be communicated widely and across all policy areas.

A new generation of innovation policy is required. Elements of this third generation policy are visible, but further development of the key ideas should be a strategic goal. The third generation innovation policy will result in innovation concepts being embedded in many policy areas. This requires much more than the issuing of pronouncements about a new policy. It will be necessary to identify and involve key stakeholders in the process, and to develop interfaces that allow for pooling of knowledge, learning from experience and evidence, and further coordination of policy initiatives. Though the third generation policy will need to be developed interactively, rather than imposed form on high, this process will require leadership and vision, with high profile and high level innovation "champions" sustaining it.

The study reaches a number of conclusions in respect of different areas of policy. First, consider two general approaches to policy reform that are currently underway – regulatory and institutional reform, and reform of governance.

Regulatory and institutional reform is an opportunity for efficient policy design processes to be introduced, and this is also an opportunity to take into account the implications of policies for innovation. It will often be appropriate for efficiency and effectiveness reasons to design measures across policy interfaces, to ensure that policies are working in the same direction. Existing approaches to Business Impact Assessment need to be further developed so as to allow all reform processes to be designed and assessed (ex ante, wherever possible) with innovation criteria to the fore. As well as stressing implications for industrial innovation, they should also allow for the impacts of reform on innovation in non-business organisations (public services, Universities, etc.) to be examined.

Systems should be installed for regular intelligence gathering, improved understanding, and benchmarking of contributions of reform to innovation. The major regulatory factors impacting innovation across all policy areas, and the relationship between different factors within and across areas, need to be identified and monitored. Governance is also important for innovation. The reform of European governance presents challenges and opportunities in this respect. Informed public opinion about broad classes of innovation must be nurtured. One element in achieving this will be the improvement of systems of public communication about RTD and innovation programmes, their design, rationale, evaluation, etc. Ways to achieve greater public involvement in decision-making as to priorities, ethics, etc. should continue to be developed and extended.

Furthermore, potential areas of social or ethical concern need to be identified and addressed. Trust in regulatory agencies must be earned (and seen to be earned) – it cannot be assumed. Thus openness and participation are important, and multiple methods to achieve these ends will need to be instituted.

In addition to these two broad themes, the study addressed a series of policy areas. In the main report these are ordered according to a rough assessment of how far they deal with factors driving innovation, generating innovative capabilities, or dealing with questions of environmental and social sustainability raised by innovation. Here we present them simply in alphabetical order. The major conclusions reached include the following:

a) Competition policy's aim of fostering greater market competition should, in general, benefit innovation. However, research indicates that there are complicated links between competition and innovation, especially in highly innovative and rapidly changing sectors.
b) In particular, technological and other innovation-related collaborations may be impeded by laws intended to restrict oligopolistic behavior and collusion. This complicated picture requires flexible design of policies around clearly stated principles that give high priority to innovation.
c) Regulatory agencies and other implementers and interpreters of policy (such as the judiciary) need to be better informed about the innovation considerations associated with decisions concerning collaborations, monopolies and mergers.

These issues are also closely entangled with matters of Intellectual Property, which also needs to be brought into the equation. Social institutions are also sources of human capital and creativity, as well as themselves being the source of many innovations and of the knowledge that underpins many more. Higher Education Institutions (HEIs) can be more entrepreneurial with respect to innovation. Policies can assist them, for instance in terms of facilitating spin-offs and interactions and collaboration with industrial and other innovators. Of course, this has to be kept compatible with the maintenance of scholarly and ethical standards. But many disincentives are inadvertently built into current institutions and regulations. These should be replaced by systems that reward individual academics for activities that link to innovation.

Equally, people with entrepreneurial and intrapreneurial experience should be enabled to contribute more to HEI research and teaching. In terms of human resources, it is important to developing individuals who combine solid disciplinary understanding with capacities to engage in multidisciplinary teamwork and to communicate across professional boundaries. Business Schools and Management Colleges, together with many other HEI courses, should be encouraged to provide high-quality training in innovation-related matters. The can be supported by such means as validation of courses and provision of suitable teaching material.

As far as *employment* is concerned, the changing nature of work, and the impacts of labour law, needs to be examined in relation to influences on innovation. This is especially true insofar as the changes affect the growing class of "knowledge workers" who are major sources of innovation. Increased mobility of such workers can raise questions of contractual restrictions – governing, for example, their use of Intellectual property and their employment in particular firms and sectors. There remains a need to develop pension, income tax, and related systems further. This should make it easier for staff to be mobile in terms of geographical location, employment, and self-employment. More generally, methods of providing support for the development of systems and procedures that reward employees for seeking innovative solutions, rather than "playing it safe", should be developed. While this is largely a matter for private initiative, public policy has a role to play in promoting awareness, good practice, and exchange of experience.

The rewards that innovative activities and "thinking outside the box" can yield should be demonstrated. High-quality material concerning innovation and entrepreneurship should also be developed for use in expanded programmes of lifelong learning.

In respect of *enterprises*, they are at the heart of successful innovation, and not just in the private sector. Entrepreneurial attitudes – even if not precisely identical motivations – underpin much innovation in public sector organisations. Support for such enterprising attitudes in general should be fostered. Small and medium sized enterprises (SMEs) will continue to remain an important focus of innovative effort, and of policy interest. The two should be brought together: innovation support facilities can be built into systems that aim at supporting SMEs in general. Support for the development of networking and innovation "clubs" is another element here. Links with HEIs and with business services that can assist SMEs' choice and implementation of innovations, and the further development and commercialisation of their own innovative ideas, should be fostered.

There is much need to continue to assist SMEs with adoption of innovations. This is especially so for those innovations that will allow them to participate on a more equal footing in the knowledge-based economy, and in some cases achieve entry to new markets and more independence from large-firm-oriented networks. Examples of support that might be specifically relevant here include for instance, web design and maintenance services for small producers and retailers. (These might best be organised on a locality basis – there are liable to be significant economies of scale and reductions in learning times associated with pooling of resources across, and services of this kind to, SMEs.)

Moreover, award systems can be good ways of promoting and diffusing knowledge of good practices, and an example here would be the introduction of awards for innovative SMEs (in "traditional" as well as "innovative" sectors), and for SME support services themselves.

Information on the drivers of innovation performance – e.g. a "benchmarking " of emerging trends in the global environment as experienced in different sectors, supply chains, regional and countries, and the responses adopted to deal with these – can contribute to building new capabilities for innovation. Enterprises and economies can build foundations for ongoing innovation and learning by competing in global value chains. Here too SMEs need support to achieve involvement appropriate to their level of technological competence.

Environmental issues are also bound to continue to grow in importance, and to receive continuing research effort in their own right. Such research should routinely include consideration of the scope for applications of innovation in support of environmental objectives. Other RTD programmes should, conversely, build into – their design, functioning, and reporting, processes – routine consideration of the contributions of these processes to enhancing sustainability. There are considerable synergies to be achieved by bringing environmental and innovation objectives closely together – and potential opportunities for distinctive EU achievements here.

Another example is in the sphere of Business Impact Assessment, where it is possible to include criteria specifically concerning positive and negative effects of regulations on environmental innovation. Environmental regulations themselves should be regularly reviewed with innovation issues in mind. For example, alternatives to mandatory and potentially technology-freezing rules should be examined. These include performance-based regulation (encouraging flexibility in finding solutions to environmental problems), and process regulation (encouraging better understanding of critical points of impact and innovations at which damage may be reduced). Information and awareness campaigns are important for alerting SMEs and less dynamic sectors to the scope for environmentally-oriented innovations.

It is important to increase recognition that environmental threats may be translated into technology strengths and market opportunities (e.g. alternative approaches to energy generation, remediation technologies, "clean" technologies).

In terms of *financial services and risk capital*, there is continued need for the development of instruments providing finance for early-stage innovation and smaller firms, with apparent gaps in availability of small-scale venture capital requiring attention. Financial support for various activities (e.g. licensing, patent investigations, etc.) also needs to be fostered. Further development of web-based financial services for SMEs is also recommended, together with appropriate awareness campaigns and support services. The financial community should be helped to acquire better intelligence about emerging areas of technological opportunity.

It also needs support in better understanding the general dynamics of innovation (e.g. time required to reach profitability, complementary assets that may be required for commercialisation, typical barriers). Better tools and standards are needed for accounting for innovation-related intangible assets and intellectual capacity in firms. Strong business participation in such a process is required, to ensure that reporting regimes and procedures benefit those regulated, as well as imposing the lowest possible new burdens on them.

Information and Communication Technologies are also essential. Policies fostering the development and use of ICTs will continue to be required as the technology itself develops, and its uses become more pervasive and manifold. Problems of access and skill persist (even if their forms change). In addition to continuing efforts to bridge skill gaps, it is thus important to continue to be vigilant against "digital divides". For example, measures may be required to ensure that SMEs are not excluded from e-markets by high entry costs, and that cheaper software and support services that are appropriate to SME business processes are available. Public bodies (local and regional agencies and HEIs as well as national governments should be encouraged to participate in the development and demonstration of innovation-oriented "knowledge management" and information systems (so as to establish standards, awareness of good practice, etc.). The scope for extended use of open source software should be further explored. The potential of ICT to be an enabling tool for many of the developments suggested for other policy areas should be exploited.

Intellectual Property Rights or IPR protection is generally seen as conducive to innovation, though most of the supporting analysis concerns the patent system. The strategies of companies with respect to patent acquisition, and, latterly, the use of copyright rules to limit the behaviour of other agents, requires careful appraisal in terms of impacts on innovation. Certainly, renewed efforts to establish common European patents are required.

However, the revisions to patent law that are under discussion require extended consultation that explicitly consider the innovation impacts of retaining or changing existing frameworks. (Examples of controversial proposals for change are modifications of the rules for dynamic sectors – e.g. shorter lifetimes of patents – and extension of the scope of patents – e.g. to cover business processes.) Similar consultations are also required to examine ways in which copyright and other rules may need to be adapted to stimulate – rather than impede – innovation. (There is rapid development of copyright law to fit it with the activities enabled by new digital media. But this urgently needs to be reviewed in the light of the innovation impacts of these developments in law and practice.) Improved advice and support should be provided to SMEs for their development and implementation of IP strategies (including negotiation with large business partners). IPR regulations and competition policy need to be jointly examined in the light of innovation trends in the knowledge-based economy.

In terms of *regional policy*, there has been considerable recognition of the regional embedded ness of much innovative activity, with clusters and systems often (not always) having a strong regional basis. The study suggests (in the case study work as much as the literature review) that it is equally important to recognise that it is most often cities and metropolitan areas that are the crucibles of innovative activity. These entities require specific attention in regional innovation policy, even though in some cases there will be bitter competition between cities to be the regional champion (while in other cases cities may be more able to co-operate). Regional innovation strategies should be helped to build more on regional distinctiveness (rather than simply identifying the same set of priorities (ICT, biotechnology, new materials ...).

It is important to recognise the significance of innovation and new technology for "traditional" sectoral activities (e.g. tourism) in cities and regions – some of these "traditional" activities are being revitalised and reshaped in the knowledge-based economy. Opportunities for linking sectoral strengths (for example, combining strengths in medical care and tourism, or in energy and environment) should be fostered.

As far as *taxation policy* is concerned, while tax removes resources that could be applied to innovation, tax rules (selective or otherwise) can be developed so as to promote innovative efforts and particular directions of innovation. On the first point, tax incentives for innovative effort are recommended. These should include but going beyond R&D, and thus methods of systematically appraising non-R&D inputs to innovation (and possibly innovation performance) should be developed. Attention should be paid not just to rewarding the level of activity, but also to encouraging continuous improvement of such effort.

There has been considerable interest in the development of environmentally-oriented taxes; these should be seen as providing an opportunity to spur innovation. Accordingly, relevant criteria should be brought into design of such taxes. Continuing attention is required concerning the question of whether a new form of taxation is required to respond to the new economic drivers of the knowledge-based economy.

Trade Policy is also an important factor. Liberalisation of trade should promote the diffusion of knowledge and innovations, though trade between countries need not spread benefits equally. Trade disputes, unfortunately, easily escalate into serious political confrontations. Thus, international efforts are required to establish mutual understanding and shared norms concerning the systematic incorporation of innovation concerns into trade negotiations and procedures. (These should include such governance issues as those connected with public acceptance of specific innovations) The EU must play a leading role in these processes, not least because there are substantial differences between European and US experience of these matters. Trade liberalisation in knowledge-based services needs coordination of rules and other practices (for example those governing professional practice and qualifications) across different countries.

Across all these diverse policy areas – and presumably across other areas of policy that have not been studied in the report such as consumer and transport policy – there are common points. The overall recommendation is that procedures for enhancing dialogue and mutual learning should be introduced to render all policies innovation friendly. This approach will support the successful implementation of the policies, and contribute to the development of third generation innovation policy.

Conclusions

As indicated above, the report as a whole and the summary presented in this chapter concerns the need and scope for reform of various policy areas in the light of the changing context of innovation, given the goals set by the Lisbon objectives. It is based upon case studies, discussions and analyses conducted in a study organised for European Commission DG Enterprise Innovation Directorate in 2001 – 2002. The overall purpose of the study was:

a) to synthesize the new approaches to innovation policy required by, and being implemented within, a knowledge-based and creative society

b) to identify and review the links between innovation policy and other policies, in particular those policies relating to the legal and regulatory framework for innovation, with a view to providing analytical pointers for further co-ordination.

In respect of the policy areas considered, it is highly likely that other policy areas would also have yielded relevant results. Furthermore, important issues such as enlargement and enlargement policy have explicitly not been taken on board in the study, since these would have increased its scope considerably.

ARW Programme

5th May 2004

Arrival and registration

19.00 – Welcome dinner

6th May 2004 (Overview of trends related to innovation)

Chair: Plamen Svetoslavov Gramatikov

08:30-09:00 – Registration

09:00-09:30 – Opening of the ARW and welcome remarks

09:30-10:00 – Presentation of participants

10:00-11:00 – Current priorities in R&D and Innovation Policy in Post-Socialist Countries: results of a survey (W Leal Filho, Germany) Presentation and Discussion

11:00-11:20 – Break

11:20-13:00 – Innovation in Europe – an overview of the state of affairs (L. Martin, UK) Paper followed by short presentations by participants

13:00-14:00 – Lunch

Chair: A. Slepoukhine

14:00-15:20 – The 6th Framework Programme: current and future trends (D. Kirrane, Belgium) – Presentation and Discussion

15:20-15:40 – Break

15:40-17:00 – Furthering innovation in Central &Eastern Europe: Polish perpectives (M. Jaskula, Poland) – Paper followed by short presentations by Anna Pobol (Belarus) and Desislava Stoilova (Bulgaria)

Evening – Reception and Dinner

7th May 2004 (Country-specific aspects of innovation)

Chair: Lynn Martin

09:00-10:50 – Moderated discussion with case studies on specific aspects of innovation in some post-socialist countries: Armenia (Boris Gabrielyan), Russia (A. Slepoukhin), Poland (Marzenna Anna Weresa), Ukraine (Olexandr Popovych, Lidiya Kavunenko)

10:50-11:20 – Break

11:20-13:00 – Innovation at the local level; trends in Bulgaria (P. Gramatikov, Bulgaria)

13:00-14:00 – Lunch

Chair: Peter Wolfmeyer

14:00-15:30 – Working groups on creating opportunities for innovation (2 groups chaired by Lynn Martin and Plamen Gramatikow)

15:30-16:00 – Break

16:00-17:00 – Groups present their results moderated by Peter Wolfmeyer

Evening – Dinner

8th May 2004 (Supporting mechanisms to innovation)

Chair: Walter Leal Filho

09:00-10:00 – CORDIS: drawing from experiences on European support to innovation (P. Wolfmeyer, Germany)

10:00-11:00 – INTAS: assisting innovation development and mobility (M. Spiesberger, Belgium)

11:00-11:20 – Break

11:20-12:20 – From a idea to a project: a step-by-step approach towards international cooperation in innovation (D. Kirrane, Belgium)

12:20-13:00 – Discussion

13:00-14:00 – Lunch

Chair: Plamen Svetoslavov Gramatikow

14:00-17:00 – Visit to a local facility

Evening – Farewell Dinner

9th May 2004

Departure of delegates

NATO Advanced Research Workshop on R&D Priorities in Innovation Policy and Financing in former Socialist countries

Blagoevrad, Bulgaria

5-8 May 2004

List of Delegates

Albania (2)

1. Dr. Bardhyl Grillo, Ph.D., Senior Inspector
Radiation Protection Commission, Ministry Of Health
80, Alexander Moisiu Str. Tirana
Albania
Telephone: +355 4 370188; Fax: +355 4 370058
E-mail: bardhig@yahoo.com

2. Mr. Dorian Jano, Ph.D. student, Aristotle University of Thessaloniki
Stil. Kiriakidi 15 G\'Feth Room A327 (Greece)
Albania
Telephone: +30 6 942499814
E-mail: janodorian@yahoo.com

Armenia (3)

3. Dr. Bardukh Gabrielyan, Ph.D., Deputy Director
Institute of Hydroecology & Ichthyology of NAS Armenia
24-D Bagramian, N 907, 375019, Yerevan
Armenia
Telephone: +3741-525424(of.); Fax: +3741-523830
E-mail: gabrielb@sci.am

4. Mr. Tigran Arzumanyan, Chief Expert, Foreign Relations Department
National Academy of Sciences of Armenia
24, Marshal Baghramyan Ave.; Yerevan 375019
Armenia
Tel: (+374 1) 520266; Fax: (+3741) 569281
E-mail: tigarz@sci.am

5. Ms. Arpine Jenderedjian; Project assistance, NGO "Orientation Union"
8 Byron Str., app. 5; Yerevan 375009
Armenia
Telephone: +(374-9)-456409; Fax: +(374-1)-523830
E-mail: arpinej@yahoo.com

Belarus (3)

6. Prof. Dr. Siarhei Zenchanka, Deputy Director, Minsk Department of the Moscow State University of Economics, Statistics and Informatics
Zheleznodorozhnaya Str., 18a; Minsk 220014
Belarus
Telephone: +375 17 222 18 66; Fax: +375 17 219 07 05
E-mail: zench@tut.by

7. Dr. Alexey Zenchenko, Ph.D., Scientific researcher
International Environmental Sakharov University
Dolgobrodskaya st., 29; Minsk 220009
Belarus
Telephone: +375 29 770 37 91; Fax: +375 17 2848929
E-mail: bems@tut.by

8. Ms. Anna Pobol, PhD Student at the Institute for Economic Research (IER), Ministry of Economy of Belarus
Guest Fellow at IAS-STS, Institute for Advanced Studies on Science, Technology and Society (IAS-STS); IAS-STS, Kopernikusgasse 9/II/12, 8010 Graz, (Austria)
Belarus
Telephone: +43 316 8126 61-23; Fax: +43 316 8126 61-11
E-mail: pobol@sts.tugraz.at; beloto@user.unibel.by

Belgium (2)

9. Dr. Manfred Spiesberger, INTAS
58/8 Ave.des arts; B-1000 Brussels
Belgium
Telephone: +32 2 549 01 44; Fax: +32 2 549 01 56
E-mail: manfred.spiesberger@intas.be
(Friday 7 May, 10.10am BXL – 1.45pm Sofia; Sunday 9 May, 4.35pm Sofia – 6.15pm BXL)

Bulgaria (9)

10. Assoc. Prof. Dr. Plamen Gramatikov, South-West University "Neofit Rilsky"
66, Ivan Mihailov Blvd.; 2700-Blagoevgrad;
Bulgaria
Telephone: +359 73 888 91 37; Fax: +359 73 88 55 16;
E-mail: psgramat@yahoo.com

11. Assist. Prof. Dr. Miroslav Daskalov, University of Plovdiv "Paisii Hilendarski"
20 Tzarevetz str.; 4002 Plovdiv
Bulgaria
Telephone: 032-642352; 098-632200; Fax: 032-642352
E-mail: invest_project@abv.bg

12. Assist. Prof. Desislava Stoilova, Ph.D. Student
Faculty of Economics; South-West University "Neofit Rilsky"
2 Krali Marko str.; 2700 Blagoevgrad
Bulgaria
Telephone: +359/87/776922; Fax: +359/73/30835
E-mail: dstoilova@abv.bg

13. Assoc. Prof. Dr. Chavdar Nikolov, Dean of the Faculty of Economics
Southwest University, Faculty of Economics
2, Krali Marko str., 2700 Blagoevgrad
Bulgaria
Telephone: +359/73/30835; Fax: +359/73/885952
E-mail: dstoilova@hotmail.com

14. Assoc. Prof. Dr. Gancho Ganchev, Senior Lecturer
Southwest University, Faculty of Economics
2 Krali Markostr.; 2700 Blagoevgrad
Bulgaria
Telephone: +359/73/30835; Fax: +359/73/885952
E-mail: dstoilova@hotmail.com

15. Prof. Dr. Mariana Maradjieva, Senior Lecturer
University of Architecture, Civil eng. and Geodesy
1 Christo Smirnenski blvd.; Sofia 1046
Bulgaria
Telephone: + 359 2 660830; Fax: +359 2 963 1796
E-mail: marmar_fhe@uacg.bg

16. Ms. Gabriela Belova, Ph.D. Student; Faculty of Law & History
South-West University "Neofit Rilky"
1, Georgi Izmirliev-Makedoncheto Sq.; 2700-Blagoevgrad
Bulgaria
Telephone: + 359 73 3 50 17; Fax: + 359 73 3 50 17
E-mail: admin@law.swu.bg

17. Assoc. Prof. Dr. Jordan Velichkov; Faculty of Law & History
South-West University "Neofit Rilky"
1, Georgi Izmirliev-Makedoncheto Sq.; 2700-Blagoevgrad
Bulgaria
Telephone: + 359 73 3 50 17; Fax: + 359 73 3 50 17
E-mail: admin@law.swu.bg

18. Assoc. Prof. Dr. Ivan Velichkov; Vice-Dean of the Faculty of Law & History
South-West University "Neofit Rilsky"
1, Georgi Izmirliev-Makedoncheto Sq.; 2700-Blagoevgrad
Bulgaria
Telephone: + 359 73 3 50 17; Fax: + 359 73 3 50 17
E-mail: admin@law.swu.bg

Germany (2)

19. Prof. Dr. Walter Leal Filho, TuTech Innovation
Harburger Schlossstrasse 6-10; 21079 Hamburg
Germany
Telephone: +49-40-76618056; Fax: +49-40-76618078
E-mail: leal@tutech.de

20. Mr. Peter Wolfmeyer, Dipl.-Kfm, ZENIT
Dohne 54; D-45468 Mülheim an der Ruhr
Germany
Telephone: +49/208/30004-31; Fax: +49/208/30004-60
E-mail: wo@zenit.de

Hungary (2)

21. Prof. Dr. Zoltan Homonnay, Head of Department of Nuclear Chemistry
Eotvos University
Pázmány P. s. 1/A; 1117 Budapest
Hungary
Telephone: 36-1-209 0555 ext. 1564; Fax: 36-1-209 0602;
E-mail: homonnay@para.chem.elte.hu

22. Prof. Dr. Gábor Dibó, Széchenyi István professor; Eotvos Loránd University
Postafiok 32; Budapest 112
Hungary, H-1518
Telephone: +36-1-372-2771; Fax: +36-1-372-2620
E-mail: dibo@szerves.chem.elte.hu

Italy (1)

23. Dr. Corrado Maria Daclon, Professor of environmental policy; University of Venice
P.O. Box 7038, 00100 Rome
Italy
Telephone: +39 335380989; Fax: +39 02700424805
E-mail: daclon@dada.it

Kazakhstan (1)

24. Dr. Marian Abisheva, Head of the Office of International Scientific and Cultural Relations
Al-Farabi Kazakh National University
71, al-Farabi ave., Almaty 480078
Kazakhstan
Telephone: +7-3272-472891 (office); Fax: +7-3272-472609
E-mail: marisha@kazsu.kz

Kyrgyzstan (1)

25. Mrs. Aigerim Toktomatova, 1 Category Specialist of the International Department,
National Academy of Sciences of the Kyrgyz Republic
Chui Prospect, Room 225, Building 265a; 720071 Bishkek
Kyrgyzstan
Telephone: + 996 312 24 35 23 (of), Fax: + 996 312 24 36 07
E-mail: ajgerimt@mail.ru

Lithuania (2)

26. Dr. Svetlana Surviliene, Senior Scientist; Institute of Chemistry
A. Gostauto 9, 01108 Vilnius
Lithuania
Telephone: +370 5 2611543; Fax: +370 5 2617018
E-mail: sveta@ktl.mii.lt

27. Ms. Vitalija Povilaityte, PhD student; Kaunas University of Technology
Kalnieciu 219-414; LT-3009 Kaunas
Lithuania
Telephone: +370 37 72 28 42; +370 37 38 87 35; Fax: +370 37 45 66 47
E-mail: vpovila@gwdg.de or Vitalija.Povilaityte@ktu.lt

FYR of Macedonia (3)

28. Prof. Dr. Konstantin Dimitrov, Faculty of Mechanical Engineering
Karpos II bb, PB 464, 1000 Skopje
FYR of Macedonia
Telephone: + 389 70 228 577; Fax: + 389 23 061 013
E-mail: kokan@ukim.edu.mk

29. Prof. Dr. Slave Armenski, Faculty of Mechanical Engineering
Karpos II bb, PB 464, 1000 Skopje
FYR of Macedonia
Telephone: + 389 70 228 577; Fax: + 389 23 061 013
E-mail: armen@ereb1.mf.ukim.edu.mk

30. Prof. Dr. Mile Dimitrovski, Faculty of Mechanical Engineering
Karpos II bb, PB 464, 1000 Skopje
FYR of Macedonia
Telephone: + 389 70 228 461; Fax: + 389 23 061 013
E-mail: dimil@ereb1.mf.ukim.edu.mk

Moldova (1)

31. Dr. Gheorghe Cuciureanu, Senior officer
Supreme Council for Science and Technological Development
180 Stefan cel Mare street, of.1208; Chisinau, 2004
Moldova
Telephone: +373 22 232563; Fax: +373 22 232873
E-mail: cuciur_scstd@yahoo.com

Poland (5)

32. Apl. Prof. Marian JASKULA, Head of the Electrochemistry Group
Faculty of Chemistry, Jagiellonian University
3 Ingardena Str., 30-060 Cracow
Poland
Telephone: +48 (0) 12 633-63-77 ext.2269; Fax: +48 (0) 12 634-05-15
E-mail: jaskula@chemia.uj.edu.pl

33. Dr. Piotr Przewrocki, Polish Academy of Sciences
Wybickiego 731-261, Krakow
Poland
Telephone: +48 502 281 932; Fax: +48 12 632 22 45;
E-mail: przewro@min-pan.krakow.pl

34. Assist. Prof. Dr. Marzenna Weresa, Warsaw School of Economics
World Economy Research Institute, Warsaw School of Economics
al. Niepodleglosci 162; 02-554 Warsaw
Poland
Telephone: +48 605 280 989; Fax: +48 22 848 91 32
E-mail: mweres@sgh.waw.pl

35. Dr. Alina Sionkowska, Lecturer/researcher; Faculty of Chemistry
Nicolaus Copernicus University
Gagarin 7 street, 87-100 Torun
Poland
Telephone: +48 56 6543150; Fax: + 48 56 6542477
E-mail: as@chem.uni.torun.pl

36. Ms. Malgozata Runevic, Researcher; Baltic States Study Center
Warsaw School of Economics, World Economy Research Institute
Rakowiecka 24 str., Warsaw 02-521
Poland
Telephone: +48 607 568 430; Fax: +48 22 848 91 32
E-mail: mrunie@sgh.waw.pl

Romania (2)

37. Prof. Dr. Ilie Simon; Academy of Economic Studies Bucharest
3, Bruxelles street, Bucharest
Romania
Telephone: +40-2-744535333
E-mail: simon@home.ro

38. Dr. Ioana Eugenia Tanasescu, Lecturer
University of Agricultural Sciences and Veterinary Medicine
1-3 Manastur str.; 3400 Cluj-Napoca
Romania
Telephone: +40.264.596384; Fax: +40.264.593792
E-mail: ioana_tanasescu@yahoo.com

Russia (4)

39. Prof. Dr. Alexander Slepukhin, Vice-Rector; Saratov State Technical University
77, Polytechnicheskaya str., 410054 Saratov
Russia
Telephone: +7 8452 506837; Fax: +7 8452 506837
E-mail: ays@sstu.ru

40. Prof. Dr. Yury Chebotarevskiy, Rector; Saratov State Technical University
77, Polytechnicheskaya str., 410054 Saratov
Russia
Telephone: +7 8452 526498; Fax: +7 8452 507740
E-mail: uvch@sstu.saratov.su

41. Prof. Dr. Vazgen Atoyan, Vice-Rector; Saratov State Technical University
77, Polytechnicheskaya str., 410054 Saratov
Russia
Telephone: +7 8452 526660; Fax: +7 8452 506740
E-mail: vra@sstu.ru

42. Assoc. Prof. Dr. Marina Butorina, Baltic State Technical University
1st Krasnoarmeyskaya Str., 1, 198005 Saint-Petersburg
Russia
Telephone: + 7 812 1101573; Fax: + 7 812 5338907
E-mail: marina_butorina@inbox.ru

Slovakia (2)

43. Ms. Tatiana Zemberyova, M.Sc., Head of International Relations
Slovak University of Technology
Vazovova 5: 812 43 Bratislava
Slovakia
Telephone: + 421 2 57294 323, 324, 325; Fax: + 421 2 57294 326
E-mail: zemberyo@rstu.vm.stuba.sk

44. Mrs. Lucia Kristoficova, International relations office
Slovak University of Technology
Vazovova 5: 812 43 Bratislava
Slovakia
Telephone: + 421 2 57294 323; Fax: + 421 2 57294 326
E-mail: kristofi@rstu.vm.stuba.sk

Turkey (1)

45. Prof. Dr. Sakir Simsek, Lecturer; Engineering Faculty, Geological Dept.
Hacettepe University
Beytepe Campus, 06532 Ankara
Turkey
Telephone: +90 312 297 77 70; Fax: + 421 2 57294 326
E-mail: ssimsek@hacettepe.edu.tr

UK (1)

46. Dr. Lynn M. Martin, Senior Academic for Innovation
University of Central England
Perry Barr G212; Birmingham B42 2SU
UK
Telephone: +44 1213317796/6366; Fax: +44 1213315222
E-mail: lynn.martin@uce.ac.uk

Ukraine (3)

47. Prof. Dr. Lidiya Kavunenko, Deputy Director on science
Centre for S&T Potential and Science History Studies (STEPS Centre)
National Academy of Ukraine
60, T.Shevchenko blvd.; 01032 Kiev-32
Ukraine
Telephone: +38-044-216-95-91; Fax: +38-044-216-95-91
E-mail: steps@carrier.kiev.ua

48. Prof., Dr. Olexandr Popovych, Head of the Laboratory; STEPS Centre
National Academy of Sciences
60, T.Shevchenko; 01032 Kiev-32
Ukraine
Telephone: +38 044 216 71 42; Fax: +38044 216 95 91
E-mail: olga_krasovska@mail.ru; steps@carrier.kiev.ua

49. Mrs. Olha Krasovska, Researcher
STEPS Center of the National Academy of Sciences of Ukraine
60, Shevchenko blvd; 01032 Kiev
Ukraine
Telephone: +38-044-216-9591; Fax: +38-044-216-9591
E-mail: steps@carrier.kiev.ua

Subject Index

Author Index